Tie of Hexosylated Hb Agglutination Counts in MOD and Stringency of Otoneural Surdity

Da Zi Shen

Tie of Hexosylated Hb Agglutination Counts in MOD and Stringency of Otoneural Surdity

First Edition August 2025

Written by Da Zi Shen

LIST OF CONTENTS

INTRODUCTION

Diabetes Mellitus is the single most important metabolic disease which can affect nearly every organ system in the body [1,2]. According to the International Diabetes Federation, the prevalence of diabetes mellitus in India is estimated to be 77 million (8.9%) and is estimated to reach 134.2 million by the year 2045. It occurs because of partial or complete reduction of insulin levels which results in increased levels of blood glucose in association with long term complications which may be vascular or neurological [3]. In addition, it affects metabolism of lipids and proteins. Among the metabolic disorders of glucose, diabetes is found commonly related to auditory disorders.

The physiological basis of type 2 diabetes mellitus is a combined form of both impairment of beta cell function, with significant raise in peripheral levels of insulin resistance at the levels of receptor and post receptor. Diabetic patients are more prone to complications of hyperglycemia, as all body cells are exposed to high levels of plasma glucose.

The complex arrangement of inner ear makes it a potential target for hyperglycaemic damage. The organ of corti cells are important structures for hearing mechanism and turns out to be the target of high glycemic levels due to micro vascular compromise and their complex structure and arrangement. The hearing loss in diabetic patients is typically described as progressive, bilateral and sensorineural type with gradual onset which usually affects higher frequencies and elderly patients[4,5]. Sensorineural hearing loss (SNHL) is a type of hearing loss, in which the root cause lies in the inner ear (cochlea and associated structures), vestibulocochlear nerve (cranial nerve VIII), or central auditory processing centres of the brain[6]. Thus screening of these patients at a higher risk of developing sensory neural hearing loss, will aid in early diagnosis and management.

Hearing impairment is defined by the World Health Organization (WHO) as a hearing loss with thresholds higher than 25dB in one or both ears. WHO states that there are approximately 63 million people who are suffering from significant auditory impairment and the prevalence being 6.3%. The degree of hearing loss is classified as mild, moderate, severe or profound[7] . Sudden sensorineural hearing loss (SSNHL) is defined as the sudden onset of unilateral sensorineural hearing loss of 30 dB over at least three contiguous audiometric frequencies[8].

Pure tone Audiometry (PTA) is described as the gold standard investigation for assessment of a hearing loss. The results of PTA are a good indicator of hearing impairment [9]. It can also be used to assess the type (conductive hearing loss, sensorineural hearing loss and mixed hearing loss) and severity of hearing loss.

Hemoglobin A1c (HbA1c) is formed by the non-enzymatic and slow glycolysis of hemoglobin in the blood; it is used to assess long-term glycemic control in diabetic patients and for diagnosis in non-diabetic patients. It indicates the risk of complications in diabetes and reflect the quality of follow-up control of patients. Although HbA1c is below 6% in individuals without diabetes, it may also exceed 10% in individuals who have uncontrolled diabetes [(10)].

This study is being done to understand the correlation between HbA1c and hearing loss by comparing the audiometric parameters. With this study we aim to bring awareness to the clinicians and general public, the need for evaluating hearing loss in diabetic patients for prevention, early detection and control of hearing loss

AIMS AND OBJECTIVES

AIM:

To evaluate the association between the severity of sensorineural hearing loss and levels of HbA1C in patients with type 2 Diabetes mellitus.

OBJECTIVES:

- To evaluate the prevalence of sensorineural hearing loss in type 2 Diabetes mellitus.
- To correlate the relation between the HbA1C titres and severity of sensorineural hearing loss.
- To evaluate the relation of severity of sensorineural hearing loss with the duration of Diabetes mellitus.

REVIEW OF LITERATURE

Diabetes mellitus is an endocrinological metabolic disorder with several complications. One of the morbid and late complications of diabetes mellitus is cochleopathy leading to irreversible sensori-neural hearing loss. Hearing loss is one of the functional disabilities which affect a person's activities. In the diabetic population, the hearing loss has been shown to be moderately high, progressive, and bilateral(11-21) . In addition to studies describing that diabetes-related hearing losses are progressive, bilateral, and involving high frequencies, there are also studies describing loss in all frequencies(22,23). The major mechanism of hearing loss in diabetes is related to micro-angiopathy of the inner ear(24-30) . The risk of cochleopathy is proportionately increased with elevated HbA1C. The values of HbA1C and the occurrence of diabetic cochleopathy are dependent on the strict glycemic control, so it is possible to have a prediction about the occurrence of cochleopathy based on the values of HbA1C, which is routinely done on a periodic basis for a diabetic individual. Since age-related hearing loss cannot be reversed, it is important to identify the preventable causes(31-33). The relationship between diabetes mellitus and the hearing loss has been assessed by previous studies.

Kathleen E. Bainbridge, Howard J. Hoffman, and Catherine C. Cowie concluded from their study on Diabetes and Hearing Impairment in the United States: Audiometric Evidence from the

National Health and Nutrition Examination Survey 1999 to 2004, that age-adjusted prevalence of low- or mid-frequency hearing impairment of mild or greater severity was 21.3% among 399 adults with diabetes as compared to 9.4% among 4741 non-diabetic adults. The association between diabetes and hearing impairment was independent of known risk factors for hearing impairment, such as noise exposure, ototoxic medication use, and smoking(34).

Abdulbari Bener, Ahmad H. A. Salahaldin, Sara M. Darwish et al inferred from their study on association between hearing loss and Type 2 Diabetes Mellitus in elderly people in a newly developed society done from the month of January 2003 upto November 2006 , that the prevalence of hearing loss was higher in men than in women along with additional evidence that diabetic mellitus type 2 has association with hearing impairments and have significant negative impact on quality of life of older persons(35).

Jianmin Ren, Peng Zhao, Li Chen et al. studied Hearing Loss in Middle-aged Subjects with Type 2 Diabetes Mellitus at Department of Otolaryngology, Qilu Hospital, Shandong University, Jinan, P.R. China in 2008 and obtained that middle-aged subjects with T2DM have subclinical hearing loss, impaired auditory brainstem response and diminished otoacoustic emissions(36).

Afaf H. Bamanie and Khaled I. Al-Noury studied prevalence of hearing loss among Saudi type 2 diabetic patients among 196 patients who attended the Otolaryngology Department at King Abdulaziz Universtiy Hospital among which Control was 87 and T2DM was 87 and identified strong relationship between T2DM and hearing loss than matched controls [(37)].

B. Viswanatha, V. Priyadarshini, M.S. Vijayashree et al. conducted a comparative study over a period of 5 years from 2008-2013 on Effect of Type II Diabetes Mellitus on 8th Cranial Nerve: Prospective Study, among 2 Groups 100 cases each: diagnosed diabetes mellitus type II in group 1 and non-diabetic controls in group 2 and concluded that there is significantly higher prevalence of hearing loss among diabetics[(38)].

Deepti Pandey, Abha Pandit, Abhay Kumar Pandey did a cross sectional study over the period of January 2011 to September 2012 in patients under outdoor care of Ashadeep Trust Hospital Indore, Central India for audio vestibular dysfunction in type2 diabetes mellitus and found hearing loss was fairly common even in mild early disease. Monitoring of such complications from the earliest, may alone, bring understanding of their pathogenesis and proper prevention and management[(39)].

Louise Ziani de David, Marcele Machado Finamor, and Ceres Buss completed a study on possible hearing implications of diabetes mellitus: a literature review, as the association between hearing loss and diabetes mellitus (type I and type II) has been debated since it was first mentioned by Jordan in 1857 and realized Diabetes mellitus is the affection most commonly related with auditory disorders, and the incidence of neurosensory hearing loss in patients with diabetes mellitus varies from 0% to 80%(40).

Karnire Nitesh Bhaskar, Sajid Chalihadan, Ravi Vaswani et al. conducted a study from October 2011 to May 2013 which included 57 cases who were diagnosed to have diabetes mellitus and 50 controls without diabetes mellitus in Yenepoya Medical College Hospital, Deralakatte, on "Clinical and Audiometric Assessment of Hearing Loss in Diabetes Mellitus" and came to a conclusion that association of SNHL with diabetes with an incidence of 78% as compared to 38% among non-diabetics.10 patients reported gradual hearing loss rest did not realize the gradual progression of hearing loss. As age and duration of diabetes increases the incidence of SNHL increases(41).

Ramlakhan Meena, Divij Sonkhya, and Nishi Sonkhya evaluated hearing loss in patients with type 2 diabetes mellitus of fifty patients as a part of hospital based observational cross sectional study carried out from February 2013 to January 2014. Study confirms the presence of SNHL in relatively young type 2 diabetes mellitus. The use of audiological tests to monitor hearing in diabetic patients should be considered as a routine procedure[42].

In a study conducted by Ooley et al. in 2017, it was revealed that in 175 patients with diabetic retinopathy and whose HbA1c value was known, there was a relationship between the hearing loss and the HbA1c value, as well as the degree of diabetic retinopathy[43] .

In a study conducted by Nagahama et al. in 2018, the hearing test results of the patient group with an HbA1c value of 7.3 and above revealed significantly higher losses at higher frequencies than other groups[44].

ANATOMY OF EAR

The ear is the organ that enables hearing and, in mammals, body balance using the vestibular system. The ears of vertebrates are placed somewhat symmetrically on either side of the head, an arrangement that aids sound localisation.

The human ear has three parts the outer, the middle and the inner ear. During the 4th week of embryonic life, about 6 tubercles appear near the 1st branchial cleft. These tubercles gradually coincide to become auricle. By 20th week, pinna achieves adult configuration. The external auditory canal forms from the 1st branchial cleft. In the 16th embryonic week, the cells multiply from below ectodermal cleft and forms the meatal plug. Recanalisation from this forms the epithelial lining in the bony meatus. External ear canal is completely formed by 28 weeks.

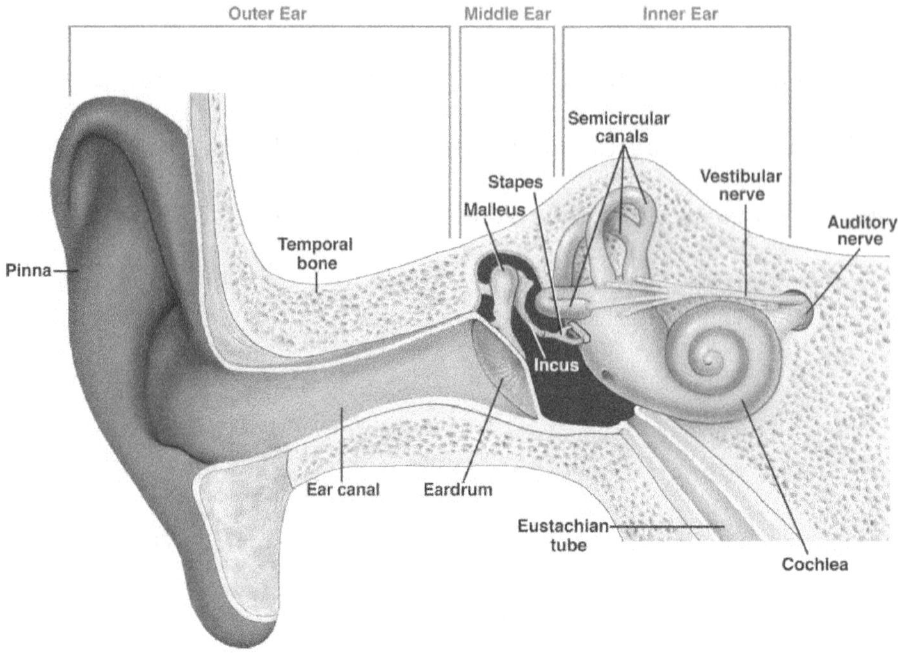

Image 1-Anatomy of ear

EXTERNAL EAR:

External ear is formed by the auricle (or pinna), external acoustic meatus (or external auditory canal) and the tympanic membrane (ear drum). The external ear collects and amplifies the sound and transmits to the middle ear. The asymmetric shape of the external auricle introduces delays in the path of sound that assist in sound localization.

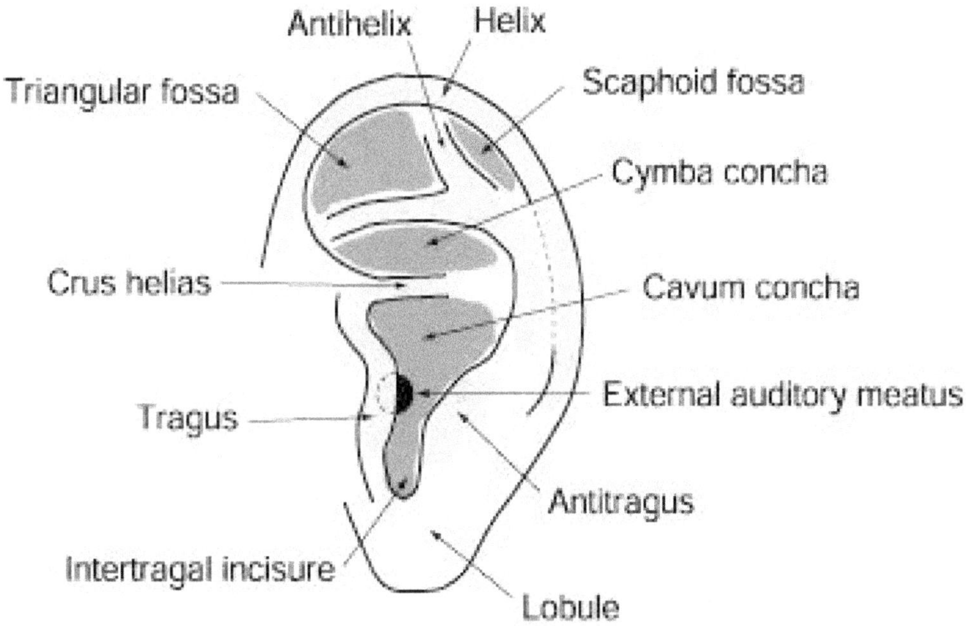

Image 2-Structure of pinna

The external acoustic meatus consists of a bony and a cartilaginous part. The outer one third of the canal is formed by cartilaginous part and the inner two third is formed by the bony part. The canal measures about 24 mm. The mandibular condyle sits anterior to the bony portion of the external acoustic meatus. The mastoid air cells sit behind the bony portion of the canal.

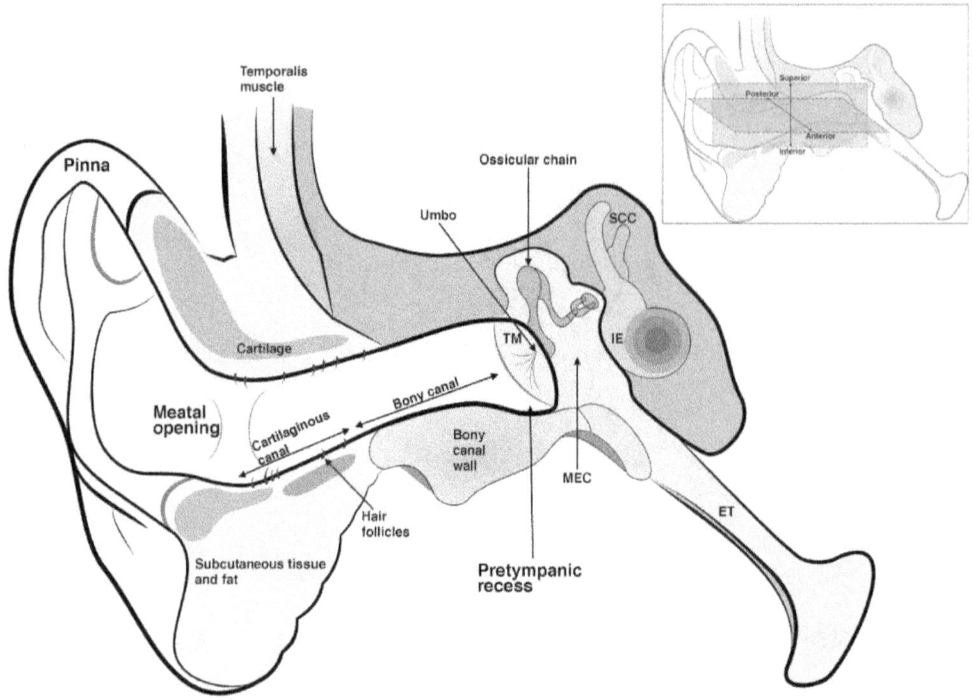

Image 3-External auditory canal

The development of tympanic membrane is from the three germinal layers. The outer epithelial layer develops from the ectoderm, the inner mucosal layer is from the endoderm and the middle fibrous layer is from the mesoderm. The tympanic membrane forms the division from the external canal and the middle ear. It is divided into two parts ,the pars tensa and the pars flaccida . Pars tensa forms most of the tympanic membrane . The central part is withdrawn inwards near the tip of the malleus and this region is called Umbo. Bright cone of light is usually found radiating from the tip of malleus extending till the periphery of the anteroinferior quadrant. Pars flaccida also known as Shrapnell's membrane is situated above the lateral process of malleus between the notch of Rivinus and the anterior and posterior malleal folds.

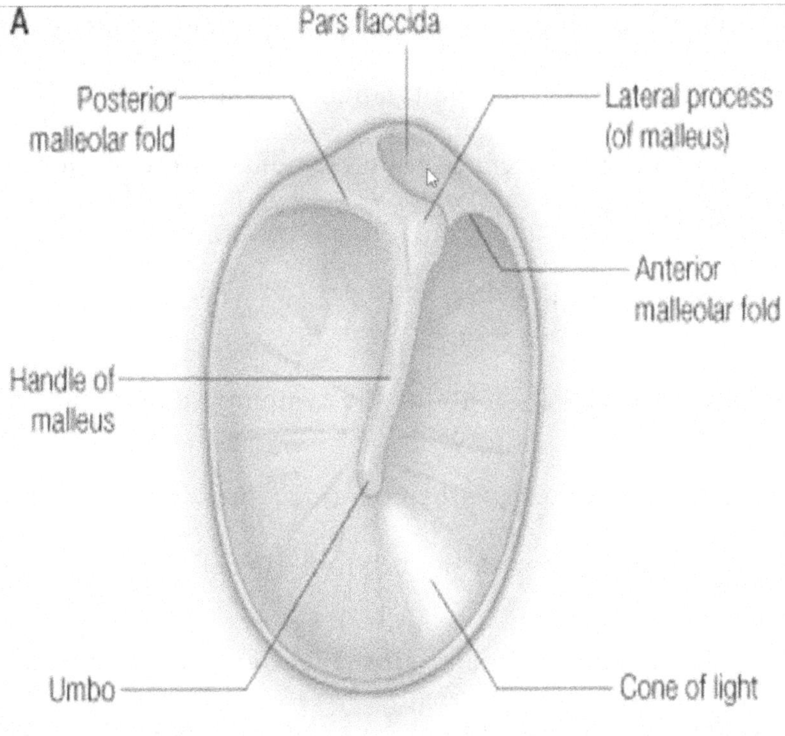

Image 4-Structure of tympanic membrane

MIDDLE EAR:

The Eustachian tube, tympanic cavity, attic, antrum and mastoid air cells forms from the endoderm in the tubotympanic recess. This is found arising from the first and second pharyngeal pouch partially. The malleus and incus develops from mesoderm of the 1st arch whereas stapes develops from the 2nd arch except the footplate and the annular ligament which is formed from the Otic capsule.

The key function of the middle ear is that of bony conduction of sound via transference of sound waves in the air from the external ear to the fluid of the inner ear. The middle ear inhabits the petrous portion of the temporal bone and is filled with air secondary to communication with the nasopharynx via the Eustachian tube. The middle ear extends from the tympanic membrane to the oval window and contains the bony conduction elements of the malleus, incus, and stapes. The walls of the tympanic cavity are complex with important associations, as follows:

- The lateral wall contains the tympanic membrane
- The posterior wall contains the mastoid antrum and communicates with the mastoid air cells
- The medial wall contains the oval window and the round window—this wall is also called the labyrinthine wall
- The anterior wall is also termed the carotid wall, because a thin plate of bone separates the carotid canal and tympanic cavity
- The roof of the tympanic cavity is the tegmental wall; it separates the middle ear from the middle cranial fossa
- The floor of the middle ear is the jugular wall; it separates the tympanic cavity from the internal jugular vein.

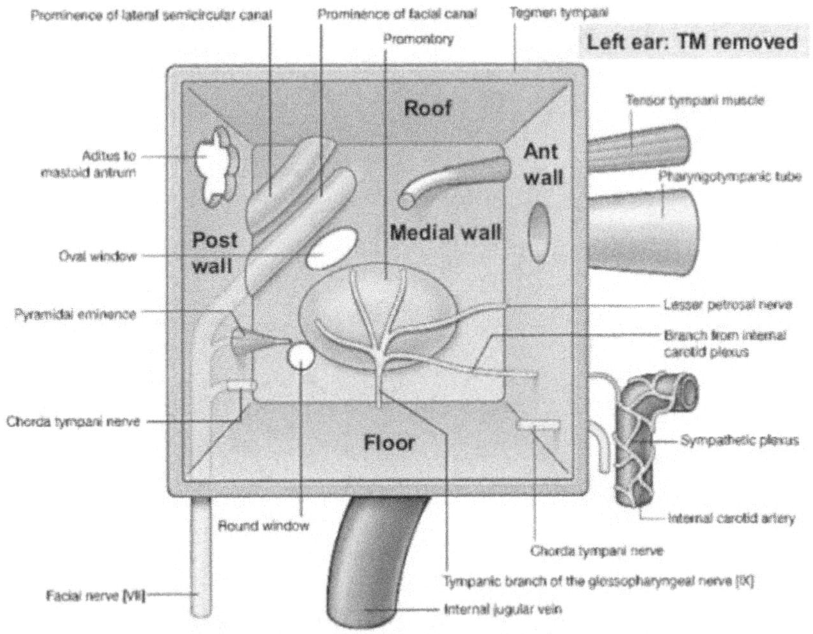

Image 5-Walls of middle ear

Middle ear ossicles are 3 in number-malleus, incus and stapes. Malleus has a head, handle, neck, a lateral process and anterior process. The head and neck of malleus lie in the attic. The incus consists of the body and short process both of which lie in the attic. The long process is attached to the head of stapes. The stapes is divided into the head, foot plate, neck, anterior and posterior crura. It is attached to oval window by the annular ligament. The ossicles direct the sound energy from the tympanic membrane into the oval window and then to the inner ear.

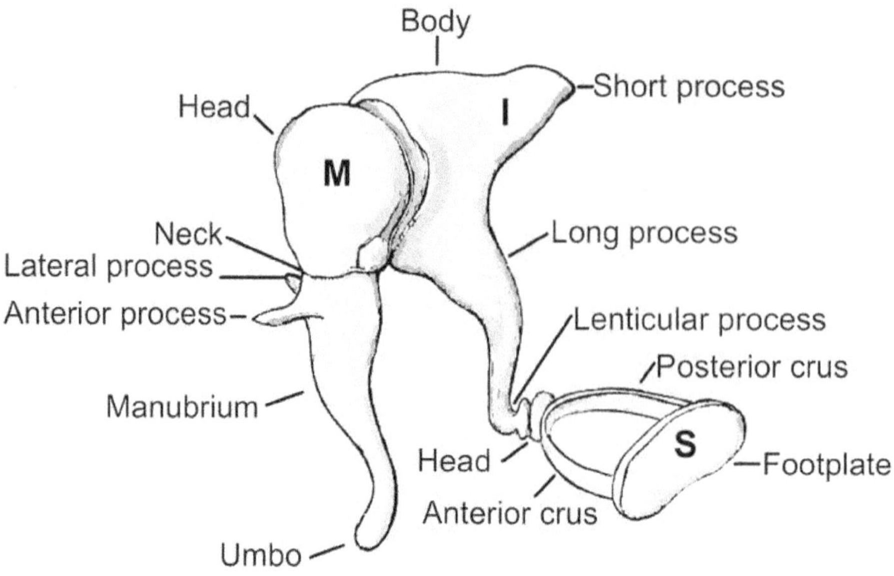

Image 6-Middle ear ossicles

ANATOMY OF INNER EAR:

Inner ear contains within the petrous part of temporal bone. It is composed of bony and membranous labyrinth .Bony Labyrinth is comprised of cochlea ,vestibule and the semi circular canals. The membranous labyrinth comprises of cochlear duct, saccule, utricle, semi circular ducts and endolymphatic duct and sac. The inner periosteal space between the bony labyrinth and membranous labyrinth is filled with perilymph which is low in potassium and rich in sodium similar to extracellular fluid, while membranous labyrinth contains endolymph which is potassium high just as intracellular fluid. These ionic composition and potentials provide the driving force for mechano transduction, hence essential for the inner ear primary function. In mammalian cochlea, organ of corti is the receptor organ, which transduces the sound stimuli to electrical signals, transmitting them to higher auditory pathway(45,46).

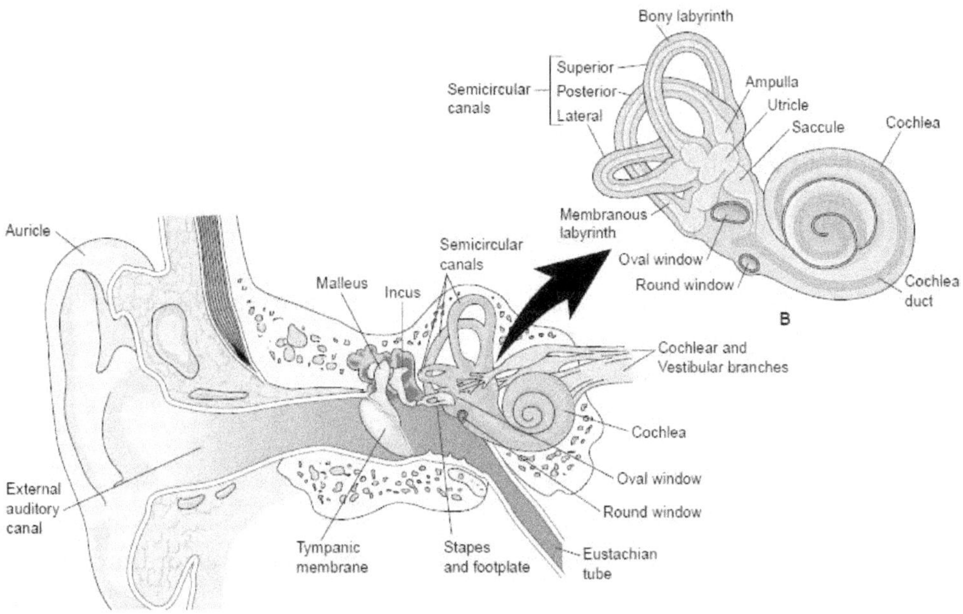

Image 7-Anatomy of inner ear

BONY LABYRINTH

It consists of three parts: the vestibule, the semicircular canals and the cochlea.

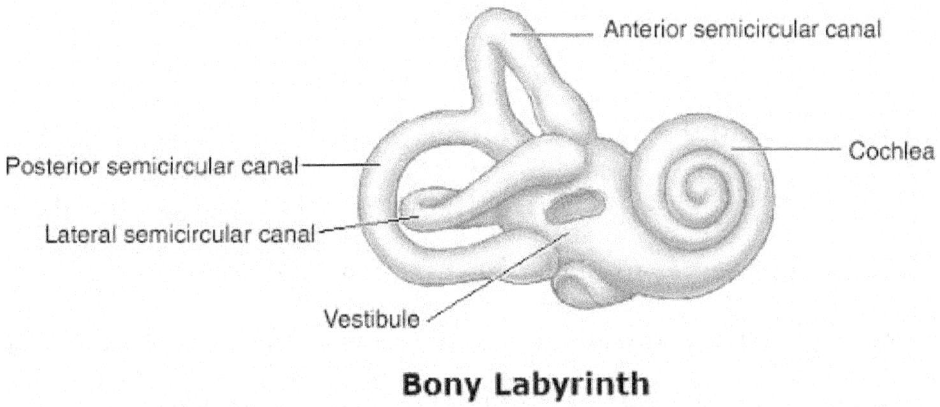

Image 8-Bony labyrinth

Vestibule.

It is the central chamber of the labyrinth. In its lateral wall lies the oval window. The inside of its medial wall presents two recesses, a spherical recess, which lodges the saccule, and an elliptical recess, which lodges the utricle. Below the elliptical recess is the opening of aqueduct of vestibule through which passes the endolymphatic duct. In the posterosuperior part of vestibule are the five openings of semicircular canals.

Semicircular canals.

They are three in number, the lateral, posterior and superior, and lie in planes at right angles to one another. Each canal has an ampullated end which opens independently into the vestibule and a non ampullated end. The non ampullated ends of posterior and superior canals unite to form a common channel called crus commune. Thus, the three canals open into the vestibule by five openings.

Cochlea.

The bony cochlea is a coiled tube making 2.5 to 2.75 turns round a central pyramid of bone called modiolus. The base of modiolus is directed towards internal acoustic meatus and transmits vessels and nerves to the cochlea. Around the modiolus and winding spirally like the thread of a screw, is a thin plate of bone called osseous spiral lamina. It divides the bony cochlea incompletely and gives attachment to the basilar membrane. The bony bulge in the medial wall of middle ear, the promontory, is due to the basal coil of the cochlea. The bony cochlea contains three compartments:

a) Scala vestibuli,
b) Scala tympani,
c) Scala media or the membranous cochlea

The scala vestibuli and scala tympani are filled with perilymph and communicate with each other at the apex of cochlea through an opening called helicotrema. Scala vestibuli is closed by the footplate of stapes which separates it from the air-filled middle ear. The scala tympani is closed by secondary tympanic membrane; it is also connected with the subarachnoid space through the aqueduct of cochlea.

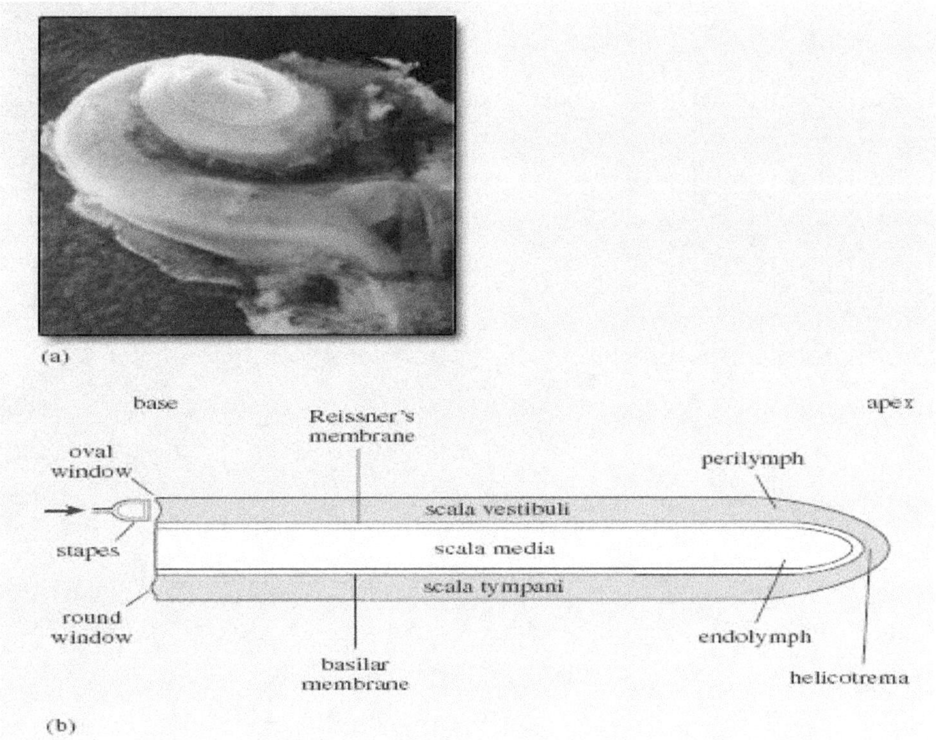

Image 9-Anatomy of cochlea

MEMBRANOUS LABYRINTH

It consists of the cochlear duct, the utricle and saccule, the three semicircular ducts, and the endolymphatic duct and sac[47] .

Cochlear duct:

Also called membranous cochlea or the scala media. It is a blind coiled tube. It appears triangular on cross-section and its three walls are formed by:

a) the basilar membrane, which supports the organ of Corti;
b) the Reissner's membrane, which separates it from the scala vestibule; and
c) the stria vascularis, which contains vascular epithelium and is concerned with secretion of endolymph.

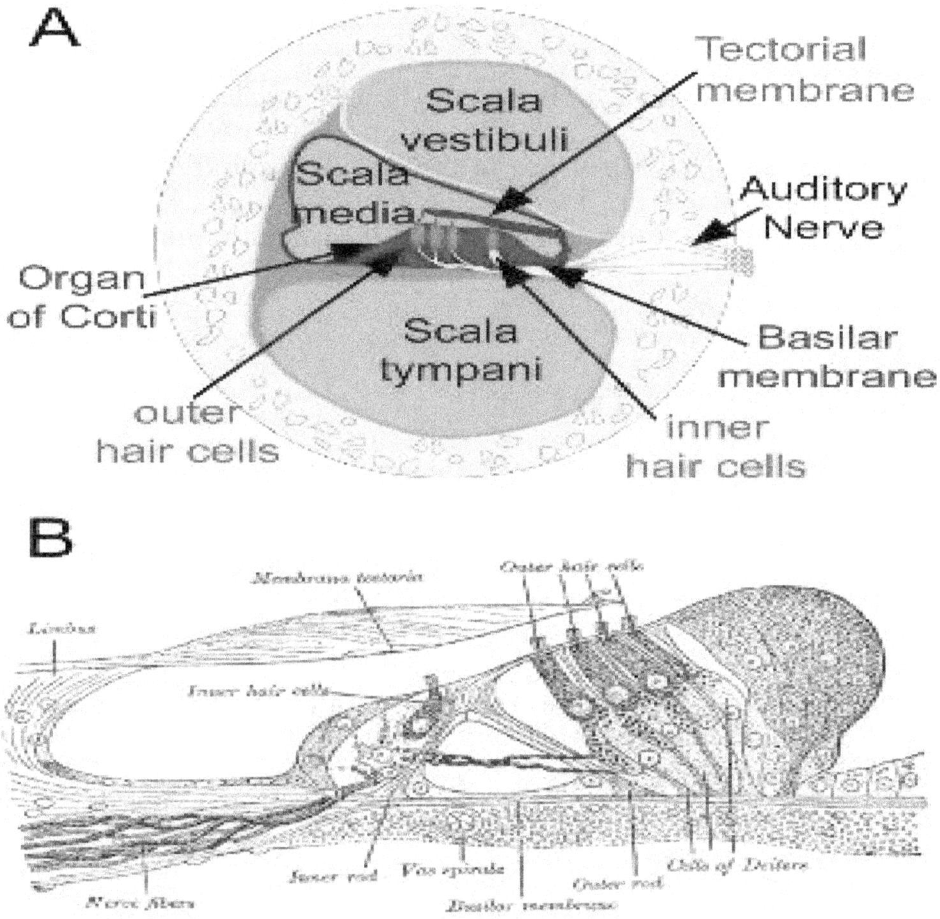

Image 10-Membranous cochlea

Utricle and saccule:

The utricle lies in the posterior part of bony vestibule. It receives the five openings of the three semicircular ducts. It is also connected to the saccule through utriculosaccular duct. The sensory epithelium of the utricle is called macula and is concerned with linear acceleration and deceleration. The saccule also lies in the bony vestibule, anterior to the utricle and opposite the stapes footplate. Its sensory epithelium is also called macula.

Semicircular ducts:

They are three in number and correspond exactly to the three bony canals. They open in the utricle. The ampullated end of each duct contains a thickened ridge of neuroepithelium called crista ampullaris.

Endolymphatic duct and sac:

Endolymphatic duct is formed by the union of two ducts, one each from the saccule and the utricle. It passes through the vestibular aqueduct. Its terminal part is dilated to form endolymphatic sac, which lies between the two layers of dura on the posterior surface of the petrous bone.

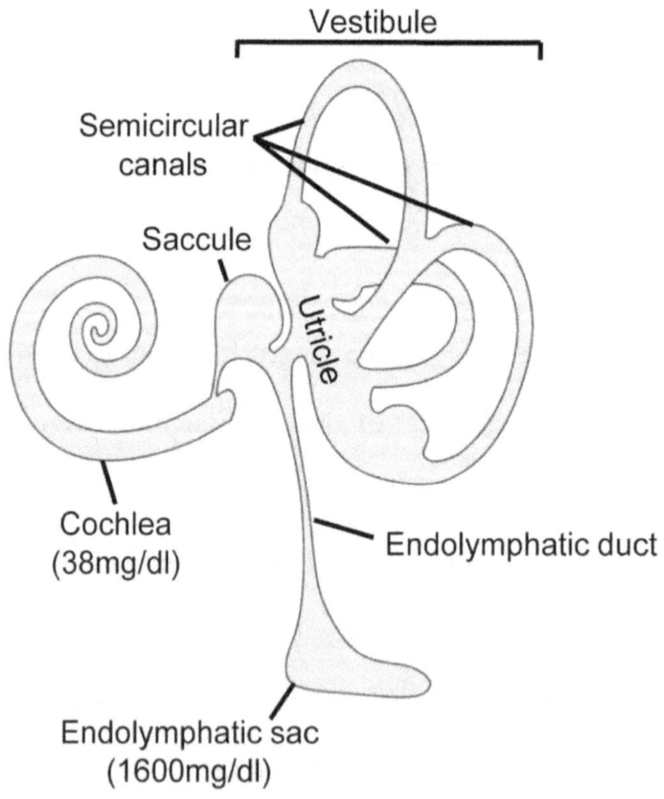

Image 11-Endolymphatic sac and duct

ULTRA STRUCTURE OF COCHLEA:

In Greek literature, Cochlea means snail. It lies anterior to vestibule. It makes 2 ¾ turns and has a height of 5mm. The cochlea's coil turns about a central core also known as modiolus. The base of the modiolus faces the inferior portion of the internal acoustic meatus. The basal coil of cochlea forms the bulge in the medial wall of middle ear called promontory [(48)].

Bony spiral lamina arises from the modiolus' edge and membranous spiral lamina extends from the edge of bony spiral lamina to the cochlea's outer wall, which divides each coil into scala vestibule (superiorly) and scala tympani (inferiorly). At the apex of the cochlea, scala vestibule and tympani communicate through the helicotrema. The footplate of stapes sits on the oval window membrane through which middle ear communicates with the inner ear. Perilymphatic fluid drains into the subarachanoid space through a bony channel called the cochlear aqueduct which is in the base of osseous spiral lamina.

Scala Media:

On the upper surface of spiral lamina, lies a spirally arranged tube called the scala media. It is triangular in cross section.

Organ of Corti, the sensory organ of hearing, resides in the membranous labyrinth[49] . The length of the basilar membrane is 35mm and width increases from base to apex. It separates scala media and scala tympani. Tectorial membrane lies over the inner and outer hair cells. Organ of Corti pillars are between inner and outer hair cells, originating from spiral lamina and basilar membrane, finally converging at the top to form the tunnel of Corti.

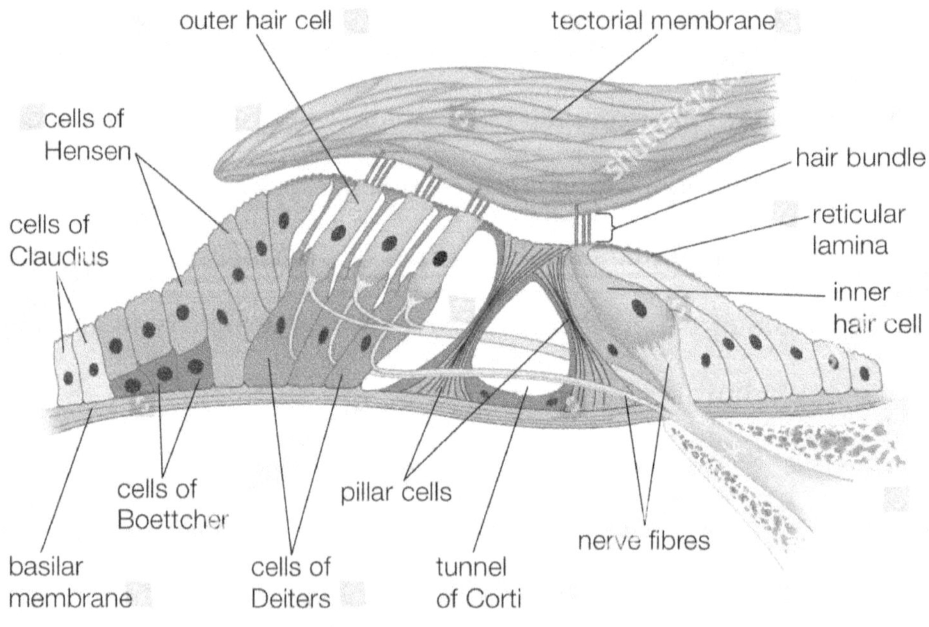

Image 12-Organ of corti

Three rows of outer hair cells lie lateral to the tunnel of Corti and is supported inferiorly by supporting Deiters cells. These cells have phalangeal process which make them project apically. Nuels space is a fluid filled space between outer hair cells and phalangeal process of deiters cells.

Cells that lie lateral to outer hair cells are: Hensen and Claudius cells. The Reticular lamina is also formed by phalangeal cells, phalangeal process of Deiter's cells and the outer hair cell's superior surface.

Scala media is separated from the scala vestibule by the Reissner's membrane, which stretches from the bony spiral lamina up to the upper part of the lateral wall of cochlear duct. 2 layers of cells comprising the Reissner's membrane are: Mesothelial layer facing the scala tympani and the endothelial layer facing the scala media.

HAIR CELLS :

They transduce mechanical sound stimulus into the electrical stimuli to stimulate auditory nerve.

INNER HAIR CELLS :

The inner hair cell is of a single row, consisting of 3500 cells, running along the inner side of sensory organ. Inner hair cells have flattened or concave apical surface and flask shaped cell body. The actin containing stereocilia are arranged according to their height. The inner hair cells comprises of two to three rows of stereocilia.

OUTER HAIR CELLS :

The outer hair cells are of 3 to 5 rows, 12000 in numbers, and has a long cylindrical shaped body. Several rows of stereocilia are present at the Apex, whose surface is flattened and the stereocilia is arranged in patterns of 'V' or 'W' shaped with bevelled tips. They contain actin filamentous core, with cross linking with plastin and espin proteins. The top of the stereocilia is in contact which may be in a floating manner with tectorial membrane. The tectorial membrane arising from spiral limbus, extends over the organ of corti and attaches in proximity to Hensen cell.

THE LATERAL WALL OF COCHLEAR DUCT :

Consists of three zones, superiorly stria vascularis, inferiorly spiral prominence and transitional zone intermediately. The stria vascularis consists of marginal, intermediate and basal cells. Marginal cells face scala media and connect with highly convoluted membranes of intermediate cells with tight junctions, which help in maintaining ionic composition of the fluids within the scala media. It also contains various enzymes and ion pumps.

Inner and outer hair cells both are overlaid by the tectorial membrane which attaches to the spiral limbus and loosely connected to supporting cells. The OHC's stereocilia embedded in the tectorial membrane enhances the cochlear frequency sensitivity and also contributes to the tonotopic cochlear organization by helping create microphonics and mechanical amplification.

Two fluid systems in the cochlea are: Perilymph present between membranous and osseous labyrinth. It has high sodium and low potassium ion concentration which makes it similar to cerebrospinal fluid. On the contrary, Endolymphatic fluid is present within the membranous labyrinth, has high potassium and low sodium ion concentration maintained by stria vascularis. The Endolymphatic sac has communication with membranous labyrinth through endolymphatic duct and vestibular aqueduct.

SPIRAL GANGLION :

The spiral ganglion located in the Rosenthal Canal contains afferent neuron cells bodies. The organ of corti release neurotransmitters that excite the dendrites. There are two types of ganglionic neurons: type I and II. Type I neurons act on inner hair cell by innervating through a converging pattern while type II neuron acts on outer hair cells in a diverging pattern of innervation. These neuronal axons project to the brain stem, maintaining the tonotopic organization of organ of corti in the afferent system also.

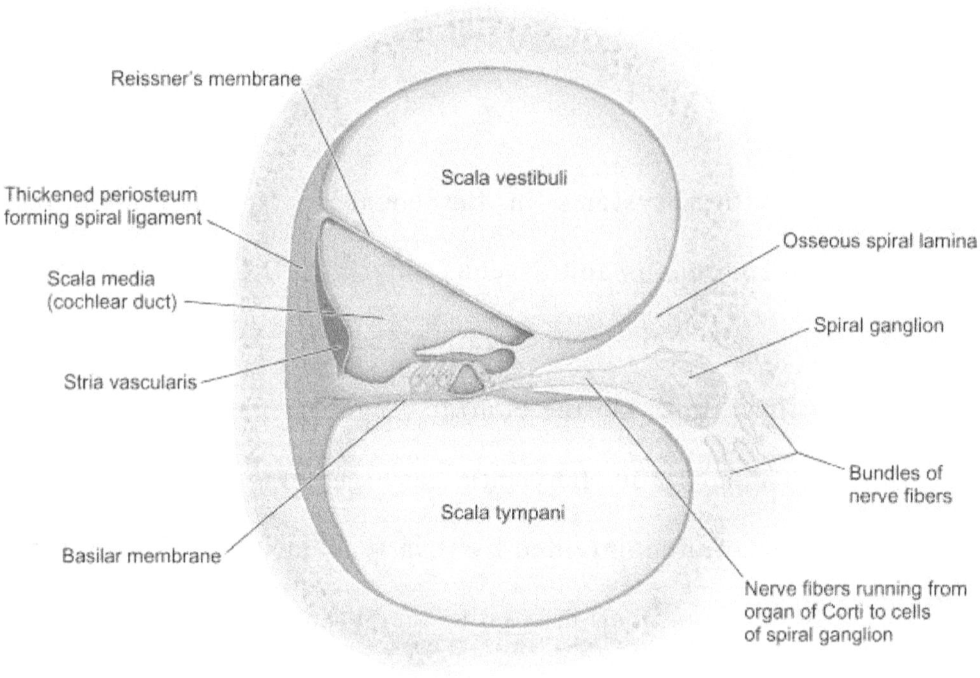

Image 13-Spiral ganglion

INNERVATIONS OF COCHLEA:

Cochlear innervations are carried out by 3 types of fibers: autonomic, afferent and efferent fibers. Autonomic fibers supply the cochlear blood vessels. Approximately 30000 auditory nerve fibers provide ascending information to central auditory pathway[50].

The auditory information transmitted through the cochlear nerve goes from inner and outer hair cells up to brain stem. Afferent fiber cell bodies forming spiral ganglion are located in the Rosenthal's canal within the modiolus. The hair cell is reached by passing through the habenulae perforatae. Majority fibers are large myelinated and project from the inner hair cells, while those unmyelinated 5-10% communicate with outer hair cells. They reach spiral canal through modiolus and project to cochlear nucleus. The maximum diameter of the cochlear nerve reaches at the base of the spiral canal.

As per the occupancy of the fibers, the low frequency are at the centre while the high frequency at the periphery of the nerve. The initial auditory processing occurs when cochlear nerve's central process reaches the cochlear nucleus, which divides into ventral and dorsal nuclei. The ventral nuclei are further subdivided into anteroventral cochlear nuclei (AVCN) and posteroventral cochlear nuclei (PVCN). The auditory nerve afferents in the AVCN terminate in the principal projection neurons of the cochlear nuclear complex and expands into very large terminal called bulb of Held.

The very low frequency fibers branch to form two end bulbs. These end bulbs contain large number of neurotransmitter vesicles and helps in rapid transmission of signal AVCN, responsible for the original frequency selectivity and sensitivity or cochlear response. The cells in these nuclei analyze the pattern of sound and determine the intensity. PVCN receives wide frequency range input and is responsible for the precise sound arrival time. The signal is then sent to the brain stem and midbrain motor nucleus and takes part in acoustic startle response. Dorsal Cochlear Nuclei complex response determines sound type. The cochlear nuclear complex auditory pathway gets divided into dorsal and ventral pathways. The dorsal pathway is a direct projection in to inferior colliculus while ventral projects to ipsilateral and contra lateral superior olivary complex, making binaural sound comparison possible.

Superior olivary complex consists of the following nuclei: 'S' shaped lateral olivary, medial olivary, medial nucleus of the trapezoid body with smaller periolivary. Medial olivary nucleus detects interaural difference in time. The 'S' shaped nuclei gets signals for excitation from the ipsilateral cochlear nuclei and inhibition from contralateral cochlear nuclei which helps detecting sound intensity difference.

Superior Olivary complex helps in sound localization. Through the lateral lemniscus the input from the brain stem auditory nuclei is projected to inferior colliculi. The two pathways emerge from the cochlear nuclear complex and join in the inferior colliculus and further analysis is made. The inferior colliculi contain a central nucleus and outer region composed of dorsal cortex and external lateral cortex. The external portion receives information from cerebral cortex.

Inferior colliculus is also involved in motor response like controlling middle ear muscle, turning head or moving eye in response to sound. The thalamus receives information from the inferior colliculi. Thalamus has medial geniculate body, posterior nucleus and reticular nucleus, which are involved in auditory function. The ventral division is organized tonotopically into low frequency layers and receives input from the central nucleus of inferior colliculus. The dorsal division is not tonotopically organized and medial division receives multimodal inputs.

From the ventral division of the medial geniculate nucleus, fibers project to Brodmann area 41 within the lateral fissure of the temporal lobe and dorsal division project to non-primary areas around A1. Most cells in A1 respond to binaural stimulation. The main function of primary auditory cortex is sound localization. Areas around A1 have complex response and it detects specific delays and simultaneous occurrence of harmonically related frequencies.

DESCENDING PATHWAYS :

The descending pathway may participate in attention level and anticipation of signals. The major one is olivocochlear feedback loop which originate in SOC and projects back to cochlea. It projects to outer hair cells and is called medial efferent system. It helps in suppression of outer hair cell mobility to make the cell less sensitive and provides protection from loud sounds. The lateral efferent system from lateral superior olivary complex supply inner hair cells which helps in sound localization and binaural comparison.

VASCULAR SUPPLY :

Arterial Supply:

Two arteries branch from common cochlear artery.

1) Main cochlear which supplies the modiolus particularly at upper basal, middle and apical coils.
2) Cochlear ramus of the vestibulocochlear artery, which supplies one-fourth of the basilar coil and modiolus.

Venous Supply:

The venous drainage of inner ear is through the veins of vestibular and cochlear aqueducts. The primary drainage of cochlea is by anterior and posterior spiral veins. The posterior vein drains the infero-medial aspect of cochlea mainly spiral ganglion, scala media and scala tympani. The anterior spiral vein drains the supero-lateral aspect mainly scala vestibuli and osseous spiral lamina. These veins enter the common modiolar vein which enters the vein of cochlear aqueduct, tributary of inferior petrosal sinus.

PHYSIOLOGY OF SOUND TRANSMISSION:

Sound is transmitted to the inner ear through the middle ear ossicles, when the sound waves strike the tympanic membrane, it increases tympanic membrane pressure in a frequency sensitive way.

An efficient middle ear impedance transformer will change the low pressure high displacement vibration of the sound waves into low displacement and high pressure vibrations.

A compression wave is developed in the inner ear fluid due to the vibration of the stapes footplate, which travels across the scala vestibuli, around the helicotrema, and out across the scala tympani toward the round window. An inward motion of the stapes causes an outward motion of the round window. This compression wave travels across the scala vestibuli. The pressure in the scala vestibuli is higher than the pressure in the scala tympani. This set up a pressure gradient, which causes the cochlear partition to vibrate. A travelling wave is set up in the basilar membrane. This movement is from base to apex. A shearing motion is developed between reticular lamina and tectorial membrane. This shearing force causes a deflection of the hair cell stereocilia. This reaches maximum at a particular place of the basilar membrane and decays.

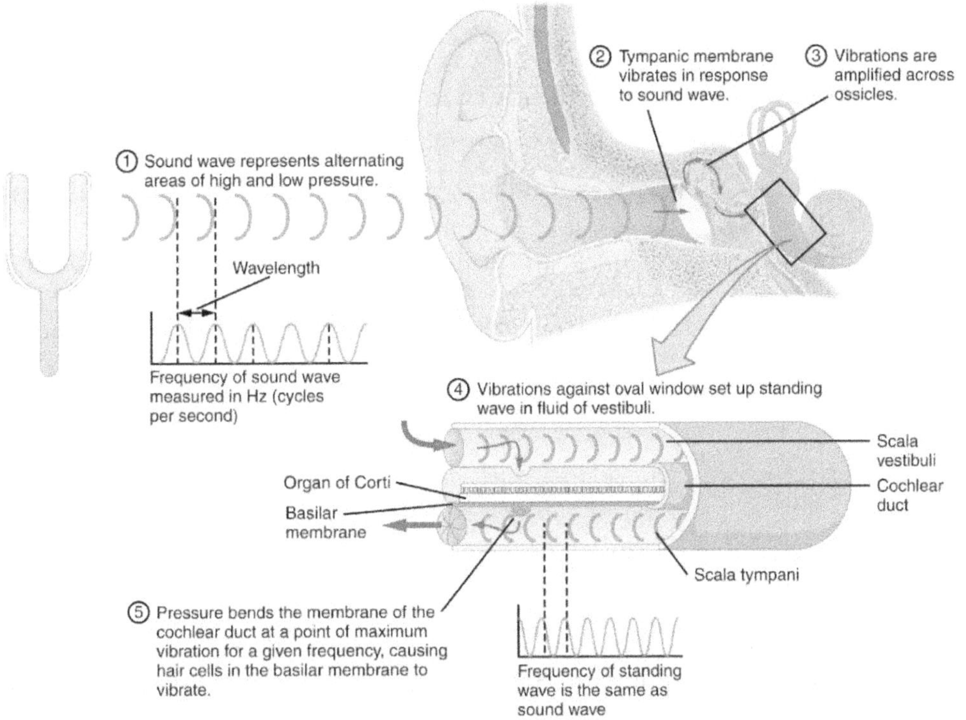

Image 14-Physiology of hearing

Physiology of hearing:

Humans can perceive sound in a range of 16 and 20000 Hertz. The signal is conducted and amplified by the outer and middle ear and finally reaches the inner ear. Here, as a function of the hair cells, the mechanical stimulus is translated to an electrical signal, which runs across the hearing pathway.

THEORIES OF HEARING

Telephonic Theory (Rutherford):

Entire cochlea responds as a whole to all frequencies instead of being activated in a plate by lace basis. Here sounds of all frequencies are transmitted as in a telephone cable & frequency analysis is performed at a higher level.

Place Theory (Helmholtz):

Basilar membrane has different segments that resonate to different frequencies.

Place-Volley Theory (Lawrence):

Combining both place and volley theory attempts to explain sound perception and transmission.

Volley Theory (Wever):

Several neurons acting as a group can fire in respond to high frequency sound even though none of them individually could do it.

Travelling Wave Theory (Georg Von Bekesy):

Sound stimulus produces a wave-like vibration of basilar membrane starting from basal turn towards apex of cochlea. It increases in amplitude as it moves until it reaches a maximum and dies off. Sound frequency is determined by point of maximum amplitude. High frequency sound causes wave with maximum amplitude near to basal turn of cochlea. Low frequency sounds have their maximum amplitude near cochlear apex.

HEARING LOSS:

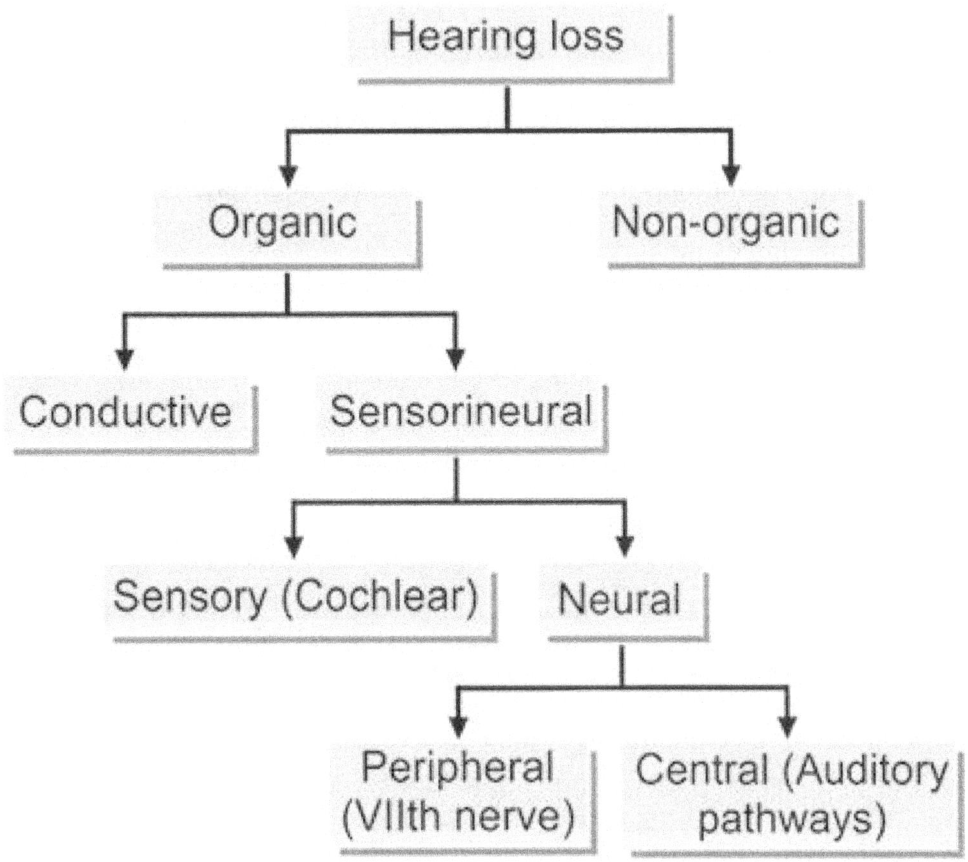

Types of hearing loss:

Based on the hearing measurements, hearing loss can be conductive, sensorineural and mixed. In the conductive type, damage of the conducting system can be located anywhere from the pinna until the stimulus reaches the hair cells.

SENSORINEURAL HEARING LOSS

In the sensorineural form, either lesion of the inner ear (cochlear) or VIIIth nerve or central auditory pathways (retrocochlear), may be present at birth (congenital) or may manifest later in life (acquired).

CONGENITAL:

It is present at birth and is the result of anomalies of the inner ear or damage to the hearing apparatus by prenatal or perinatal factors.

ACQUIRED:

Appears later in life. Genetic hearing loss may manifest late (delayed onset) and may affect only the hearing, or present as a syndrome with multi-system involvement (syndromal).

Characteristics of Sensorineural Hearing Loss in a tuning fork test are:

1) Positive Rinne test, i.e. AC > BC.
2) Weber lateralized to better ear.
3) Bone conduction reduced on ABC test.

Etiology of sensorineural hearing loss:

Sensorineural hearing loss can be present at birth or may develop during later stages of life. The cause of congenital sensorineural hearing loss remains unclear in almost half of the patients. Usually, functional failure of the hair cells can be found in these patients, but in 20% of the cases, bony malformation of the cochlea can be detected as well.

Hearing loss due to environmental factors can develop in any stage of life. Among these factors, usage of drugs with ototoxic side effects like aminoglycoside antibiotics (kanamycin, gentamicin) or diuretics (furosemide, ethacrynic acid) is very common.

Degree of hearing loss:

Hearing loss can also be classified according to its severity: Mild, moderate, severe and profound hearing loss as well as complete deafness can be distinguished. The type of rehabilitation is determined by the degree of hearing loss.

Pathogenesis of Sensorineural Hearing Loss in Diabetics:

The three main theories in the pathogenesis of sensorineural hearing loss in diabetic patients are microangiopathy of cochlear vessels, neuropathy of auditory nerve and alteration in inner ear glucose levels. The angiopathy of the inner ear leads to deafness either by diminution of transport of nutrients through thickened capillary walls or by diminution of blood flow through narrowed vasculature.

The primary diabetic neuropathy may be due to accumulation of Sorbitol within the nervous tissue, the secondary neuropathy is due to decrease in the blood flow of vasa nervosum causing secondary degeneration of eighth cranial nerve. The inner ear utilizes glucose to produce energy and changes in the glucose concentration in the inner ear may alter hearing. Recently it has been recognized that diabetics with mitochondrial DNA abnormalities are associated with sensorineural hearing loss. The pathogenesis being disruption of energy production which thus compromises tissues with high metabolic energy requirements like labyrinth of the ear.

Histopathological changes seen in inner ear in diabetes mellitus:

Microangiopathic changes with PAS positive precipitates in stria vascularis, internal auditory artery, modiolus, vasa nervosum of eighth nerve and spiral ligament. There was also demonstration of hemorrhage in endolymph and perilymph along with loss of hair cells, atrophy of spiral ganglion demyelination of eighth nerve and degenerative changes in brain stem and cerebellum.

NEUROPATHY IN DIABETES MELLITUS

Neuropathy affects around 50% of diabetics during their lifetime exhibiting as polyneuropathy, mononeuropathy or autonomic neuropathy. The development and severity of neuropathy is determined by the duration of diabetes and degree of metabolic control. Diabetes mellitus affects both myelinated and non-myelinated nerve fibers.

In diabetics of short duration, institution of insulin therapy improves the impaired motor conduction velocity. Loss of myelin, corresponding to segments of nerve underlying individual Schwann cells has been found in diabetes. There is also a possibility that diabetic neuropathy affects primarily either the cell body or axon and that segmental demyelination is a secondary event.

The structural changes in the neurons of peripheral nervous system due to age related pathologies show loss of cells and axons, decrease in size of neurons and axons and pre-degenerative changes.

There is threefold increase in prevalence of parasympathetic and sympathetic neuropathy in patients with diabetic neuropathy, ten years after the initial diagnosis of the same. The proposed abnormalities implicated in the pathogenesis of diabetes neuropathy are:

1. Basement membrane thickening and reduplication.
2. The endothelial cell swelling and pericyte degeneration
3. Occlusive platelet thrombi
4. Closed capillaries
5. Multifocal ischemic proximal nerve lesions
6. Epineural vessel atherosclerosis
7. Increased oxygen and free radical activity
8. Reduced endothelial nitric oxide activity
9. Nerve hypoxia
10. Advanced glycation of vessel wall

PURE TONE AUDIOMETRY:

In pure tone audiometry, we test the hearing sensitivity of a subject only for pure tone sounds. But the sounds that we hear in nature are complex sounds and not pure tone sounds. A complex sound is composed of sounds of various frequencies and intensities. However, all complex sounds are nothing but a mixture of different pure tone sounds and can be mathematically broken into their waves of different frequencies by a process known as Fourier's Analysis.

Hence, by estimating the pure tone hearing threshold in a number of frequencies as we do in PTA, we can have a fairly accurate idea of the hearing sensitivity of a subject for complex sounds.

The pure tone audiometry test, in addition to providing a quantitative measurement of the subject's hearing, also gives a scientific credibility to the results of the clinical tests. Moreover, the audiometry test establishes a baseline for any changes which may occur as a result of treatment or due to the natural progression of the disorder. The aims of performing the PTA test are to ascertain:

a) Whether the subject has any definite hearing loss
b) Whether the hearing loss is conductive/ sensorineural/mixed
c) If sensorineural, then whether it is cochlear or retrocochlear
d) The degree of hearing dysfunction.

Though the PTA test does not determine the exact pathology of the disorder, nevertheless, by broadly classifying the deafness into three categories, i.e., conductive/sensorineural/mixed, it does help to limit the number of possibilities in the diagnostic work-up.

The result, when plotted graphically, is called a pure tone audiogram. The instrument used for this is an electronic device called pure tone audiometer. Pure tone audiometer consists of an audio-oscillator which generates pure tone sounds of various frequencies usually at regular steps of 125, 250, 500, 1000, 1500, 2000, 3000, 4000, 6000 and 8000 Hz. The tones are attenuated by an attenuator dial which is marked in decibels and graduated in 5dB steps from -10 to 110dB. Most modern sophisticated audiometers however have graduations in steps of 2 dB and 1 dB also in addition to the 5 dB steps, which is sufficient for clinical audiometry tests.

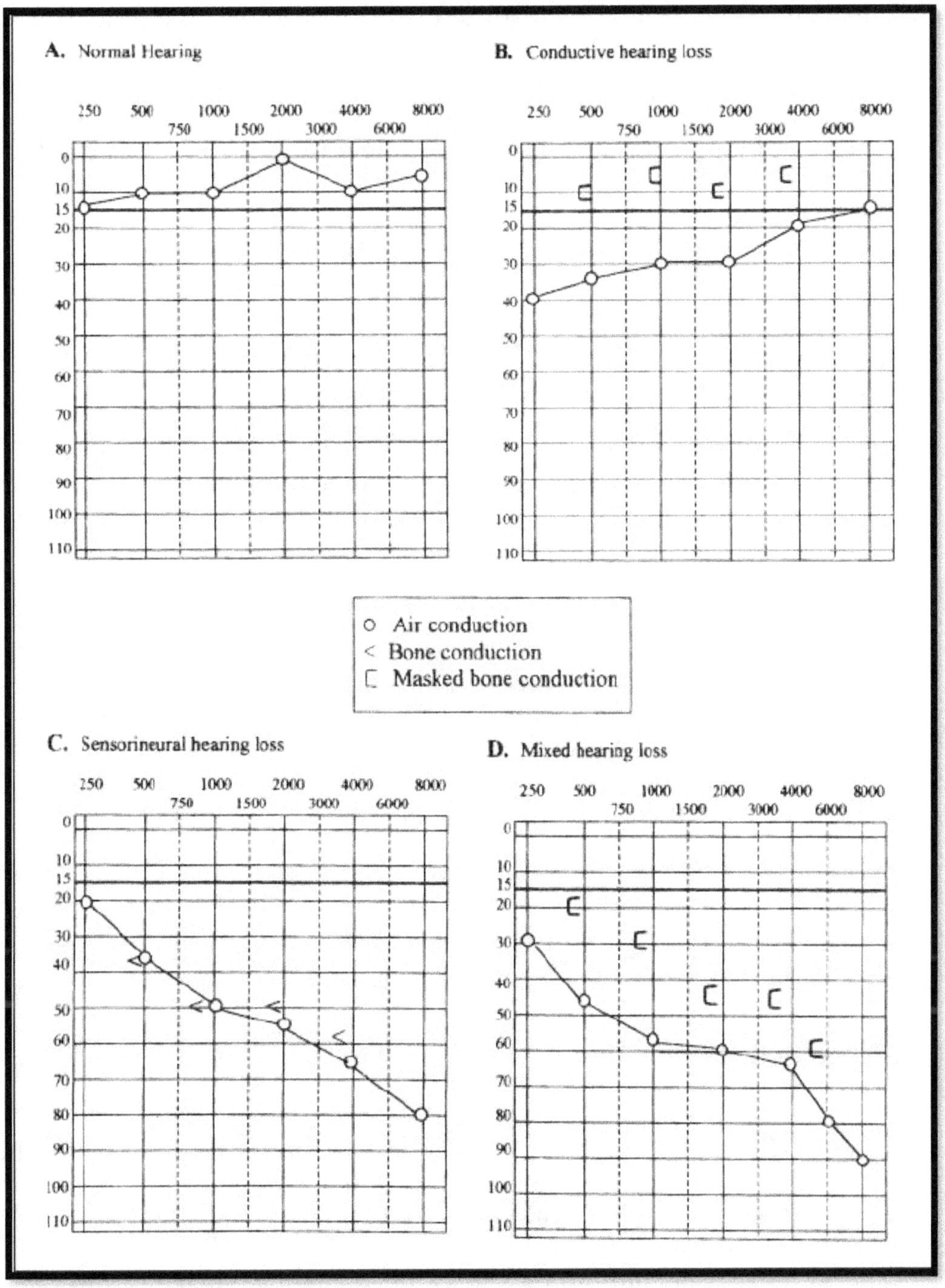

Image 17- Pure tone audiogram

The audiometer is connected to standard and specified earphones or to a bone conduction vibrator through which the sound is presented to the subject's ears. The ear phones which are placed over the ears or may be intra-aural insert ear-phones which are deep inside the external auditory meatus. The audiometer is operated by means of a noiseless switch called interrupter which can introduce or interrupt a tone.

LIMITATIONS AND SHORTCOMINGS OF PTA

1. Audiograms are often inaccurate due to improper technique, test condition, test instrument, examiner.
2. Subjective test.
3. Does not assess other salient features of hearing as it assesses hearing at threshold only.
4. Bone conduction test has challenges in giving out the true sensorineural reserve.

DIAGNOSIS

- History of preexisting diabetes mellitus for a long term and no history of other otological problems.
- Pure tone audiogram shows sensorineural hearing loss.

MATERIALS AND METHODS

This study is a prospective observational study conducted in the Department of Otorhinolaryngology, Saveetha Medical College and Hospital, Kancheepuram from November 2020- March 2022. This study was done to assess the audiological characteristics in diabetic patients and to evaluate the association between the severity of sensorineural hearing loss and HbA1C levels in them. The study was approved by the Institutional Review Board, Ethical Committee of Saveetha University and the study was conducted in accordance with ethical standards established by Declaration of Helsinki (2000). All patients were explained about the study and a written and informed consent was obtained prior to participation. A total number of 100 patients were included in this study (n=100). Convenience sampling was done and diabetic patients reporting to the hospital were included.

Study design :

Hospital based prospective observational study

Inclusion criteria :

- Patients with type 2 Diabetes mellitus
- Age group:30 -55 years

Exclusion criteria :

- Patients with history of hearing loss prior to the onset of diabetes
- Patients with history of ear discharge, any other otologic lesion, oto-surgical intervention, ototoxic drug intake
- Chronic noise exposure
- Smokers
- Head injury patients
- Patients with other co-morbidities affecting hearing
- Age less than 30 and above 55 years

Research design:

Among the patients who were included in the study as per the above mentioned criteria, a detailed history about hearing loss (duration, onset, associated symptoms) and diabetes (duration, and treatment) will be obtained. They will be then subjected to a detailed ENT and complete otoscopic examination including tuning fork tests.

Patients who fit the criteria for the study will undergo further assessment

1) Audiometric assessment-Pure tone audiometry (PTA).
2) Blood investigation-FBS, PPBS and glycosylated hemoglobin HbA1c.

Patients underwent imaging to rule out other neural causes of hearing loss and space occupying lesions.

Pure tone audiometry was performed by the department audiologist in sound treated room after necessary clinical examination for hearing loss was undertaken for all patients. PTA was done at varied frequencies of 250Hz, 500 Hz, 1 KHz, 4 KHz, 8 KHz.

The hearing threshold was graded as per World Health Organization (WHO) guidelines.

	Grades of impairment	**PTA-hearing threshold**
0	No impairment	<25 dBHL
1	Slight impairment	26-40 dBHL
2	Moderate impairment	41-60 dBHL
3	Severe impairment	61-80 dBHL
4	Profound hearing loss	>80 dBHL

The patients will be divided into 3 groups based on their HbA1C levels

- HbA1C level<7%: Excellent control
- HbA1C level 7-8%: Good control
- HbA1C level>8%: Poor control

The HbA1C levels will be correlated with the degree of sensorineural hearing loss.

Statistical analysis:

p value was calculated using t test and chi square test.A 'p' value of <0.05 was considered as statistically significant.

Sampling Technique:

Non probability convenience method

Statistical analysis:

Chi – squared test and Descriptive statistics

Data Collection Methods:

Periodicity:

November 2020 to November 2021

Setting:

Saveetha medical college and Hospital, Kancheepuram

- Department of ENT
- Department of Speech and Audiology.

Benefits and Risks of the Study

Appropriate early diagnosis of causative factors for SNHL in diabetics and to ensure prompt and effective management to avoid or minimize the occurrence of complications.

No risks so far have been reported in prior studies.

RESULTS AND ANALYSIS

109 patients were included in this study as per inclusion criteria. 9 patients were excluded from the study due to smoking and pre-existing hearing loss. The 100 patients were analysed as per demographic details and their hearing loss and HbA1c levels were correlated and the results are shown below.

GENDER WISE DISTRIBUTION:

Among 100 patients, 51 were male (51%) and 49 were female (49%).

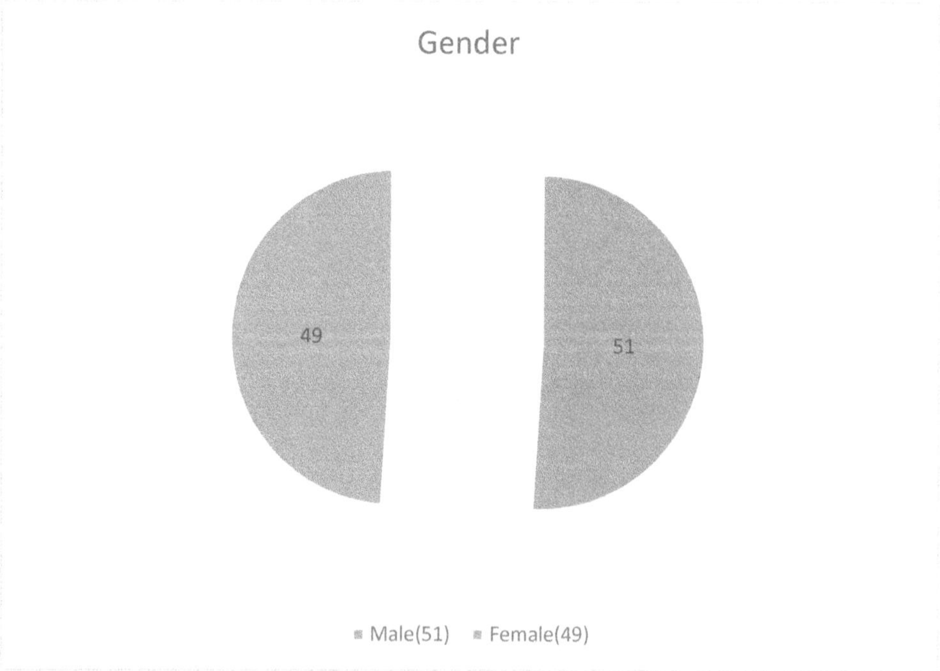

AGE WISE DISTRIBUTION:

The study population was divided into 5 age groups. The age group of 31-35 years included 6 patients, 36-40 years included 14 patients, 41-45 years included 24 patients, 46-50 years included 33 patients and 51-55 years included 23 patients. The mean age group being 45.96 years and the maximum number of patients being in the age group of 46-50 years.

AGE GROUP	NUMBER OF PATIENTS
31-35 YEARS	6
36-40 YEARS	14
41-45 YEARS	24
46-50 YEARS	33
51-55 YEARS	23

PREVALENCE OF HEARING LOSS:

Among the study population, 59 patients (59%) had hearing loss and 41 patients (41%) didn't complaint of hearing loss.

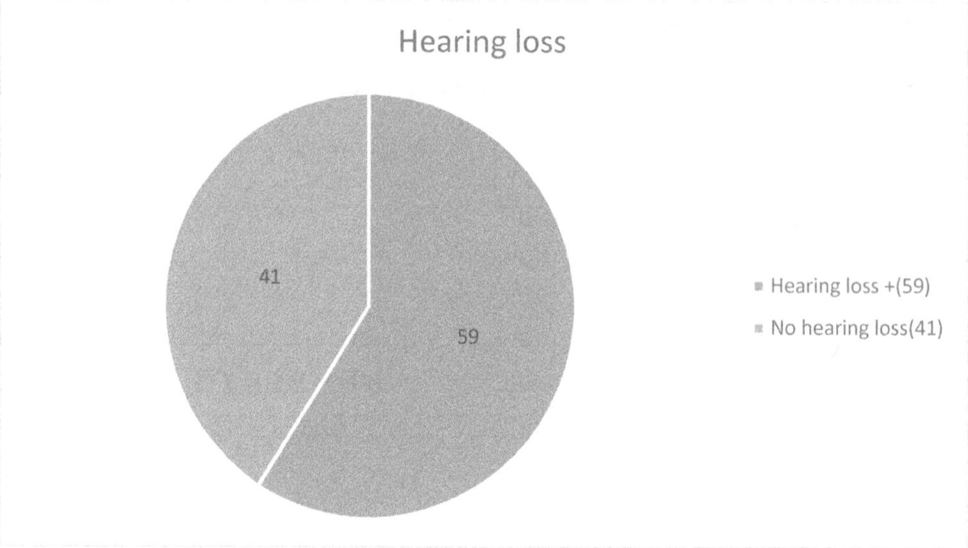

SIDE OF HEARING LOSS:

Among the 59 patients with hearing loss ,37 patients had bilateral hearing loss (63%) ,12 patients had left sided hearing loss (20%) and 10 patients had right sided hearing loss (17%).

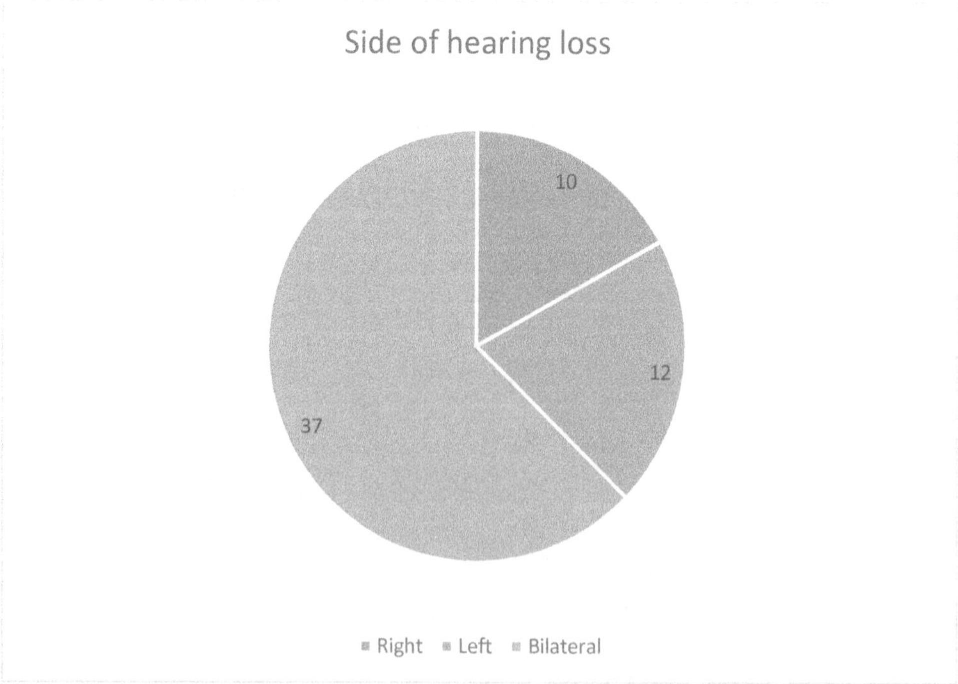

TYPE OF HEARING LOSS:

Among the study group, 5 patients had conductive hearing loss, 7 patients had mixed hearing loss, 47 patients had sensorineural hearing loss and hearing was within normal limits for 41 patients.

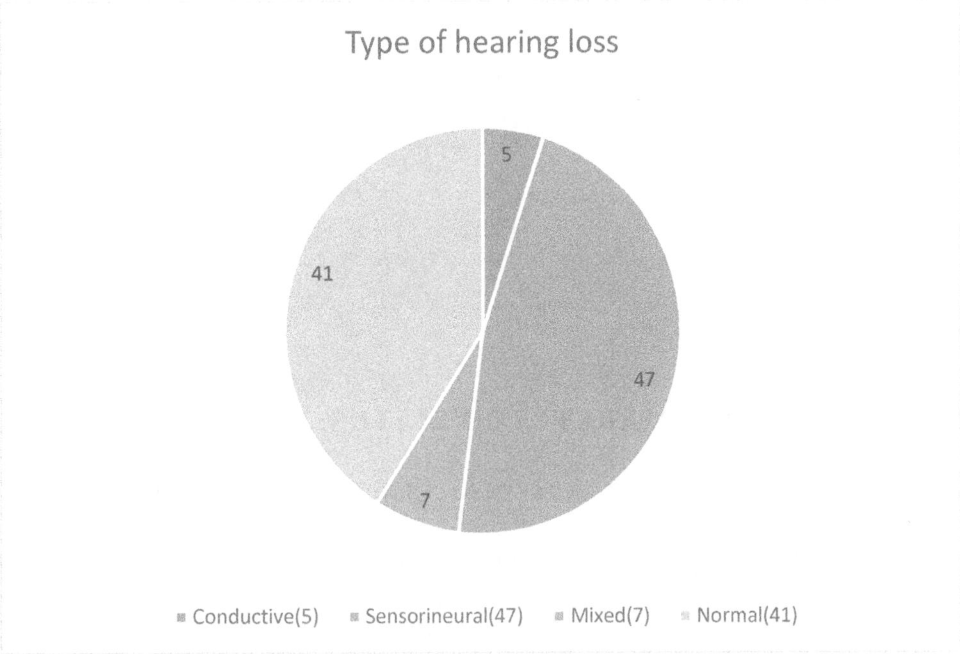

DEGREE OF HEARING LOSS:

Among the study population, 13 patients had mild hearing impairment, 35 patients had moderate hearing impairment, 11 patients had severe hearing impairment and hearing was within normal limits for 41 patients.

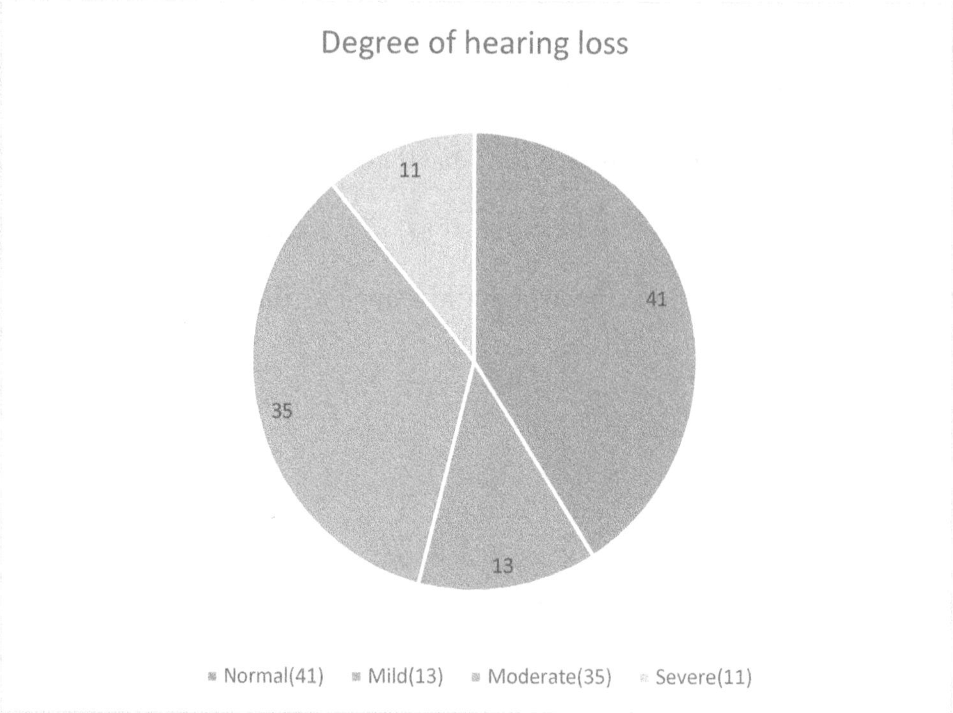

DURATION OF DIABETES MELLITUS:

All 100 patients who participated in this study are diabetic. Among the study group, 24 patients had diabetes mellitus for a time span of <=5 years, 34 patients had diabetes mellitus for a duration of 5-10 years and 42 patients had diabetes mellitus for a duration of more than 10 years.

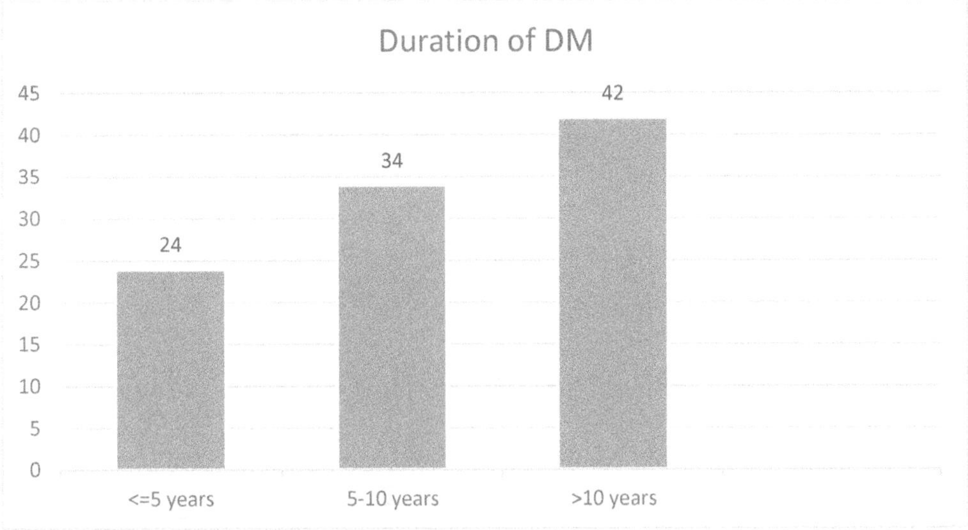

PREVALENCE OF HEARING LOSS WITH DURATION OF DIABETES MELLITUS:

Among the 24 patients whose duration of diabetes mellitus is <=5 years, 4 patients (17%) had hearing loss and 20 patients (83%) had no hearing loss. Among the 34 patients who had diabetes mellitus for 5-10 years, 19 patients (56%) had hearing loss and 15 patients (44%) had no hearing loss. Among the 42 patients who had diabetes mellitus >10 years ,36 patients (86%) had hearing loss and 6 patients (14%) had no hearing loss. The p value is <0.05, making it statistically significant.

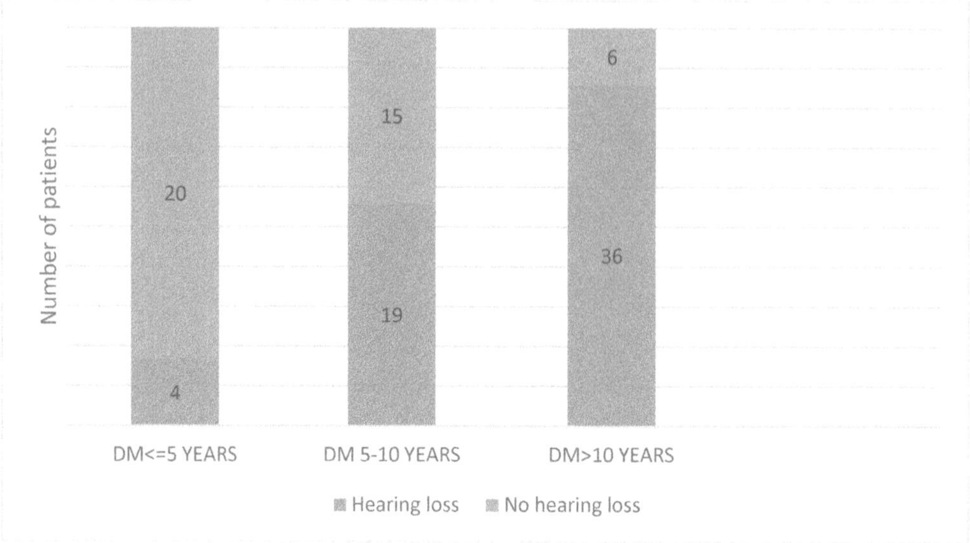

COMPARISON OF TYPE OF HEARING LOSS WITH DURATION OF DIABETES MELLITUS:

Among the 24 patients whose duration of diabetes mellitus is <=5 years, 1 patient (4%) had conductive hearing loss, 2 patients (8%) had sensorineural hearing loss, 1 patient (4%) had mixed hearing loss and 20 patients (83%) had no hearing loss. Among the 34 patients who had diabetes mellitus for 5-10 years, 2 patients (6%) had conductive hearing loss ,14 patients (41%) had sensorineural hearing loss and 3 patients (9%) had mixed hearing loss and 15 patients (44%) had no hearing loss. Among the 42 patients who had diabetes mellitus >10 years ,2 patients (5%) had conductive hearing loss ,31 patients (74%) had sensorineural hearing loss ,3 patients (7%) had mixed hearing loss and 6 patients (14%) had no hearing loss.

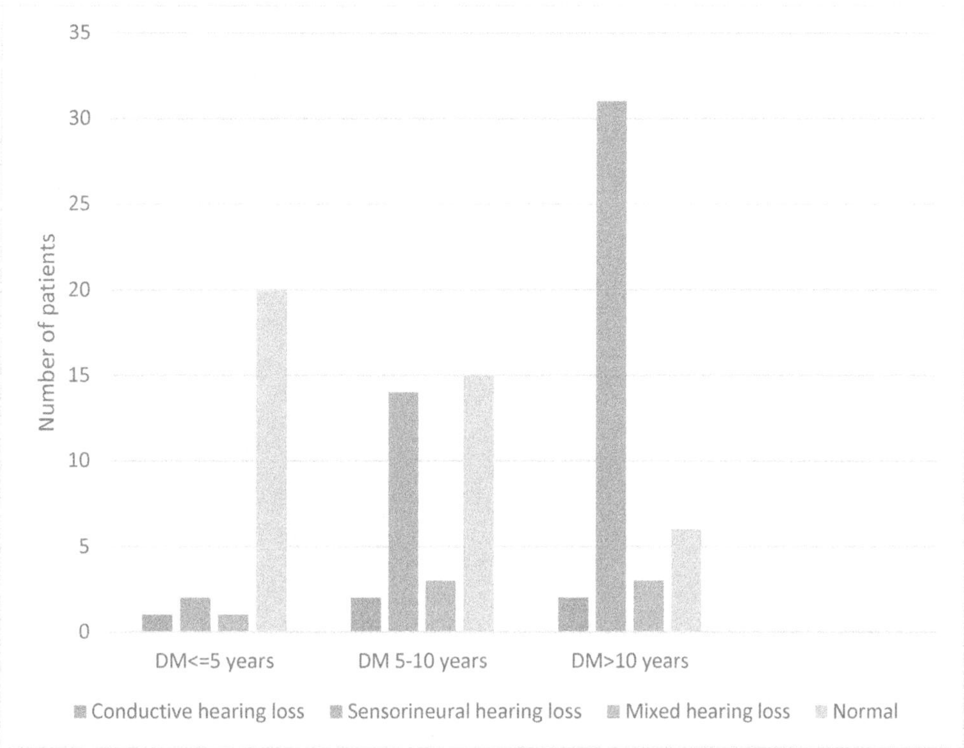

COMPARISON OF SEVERITY OF HEARING LOSS WITH THE DURATION OF DIABETES MELLITUS:

Among the 24 patients who had DM<=5 years ,20 patients (83%) had hearing within normal limits ,3 patients (13%) had mild hearing loss ,1 patient (4%) had moderate hearing loss. Among the 34 patients who had DM for 5-10 years ,8 patients (24%) had mild hearing loss ,8 patients (24%) had moderate hearing loss and 3 patients (8%) had severe hearing loss and 15 patients (44%) had no hearing loss. Among the 42 patients who had DM>10 years ,2 patients(5%) had mild hearing loss ,26 patients(62%) had moderate hearing loss ,8 patients(19%) had severe hearing loss and 6 patients(14%) had no hearing loss .The p value is <0.05 and hence, making it statistically significant.

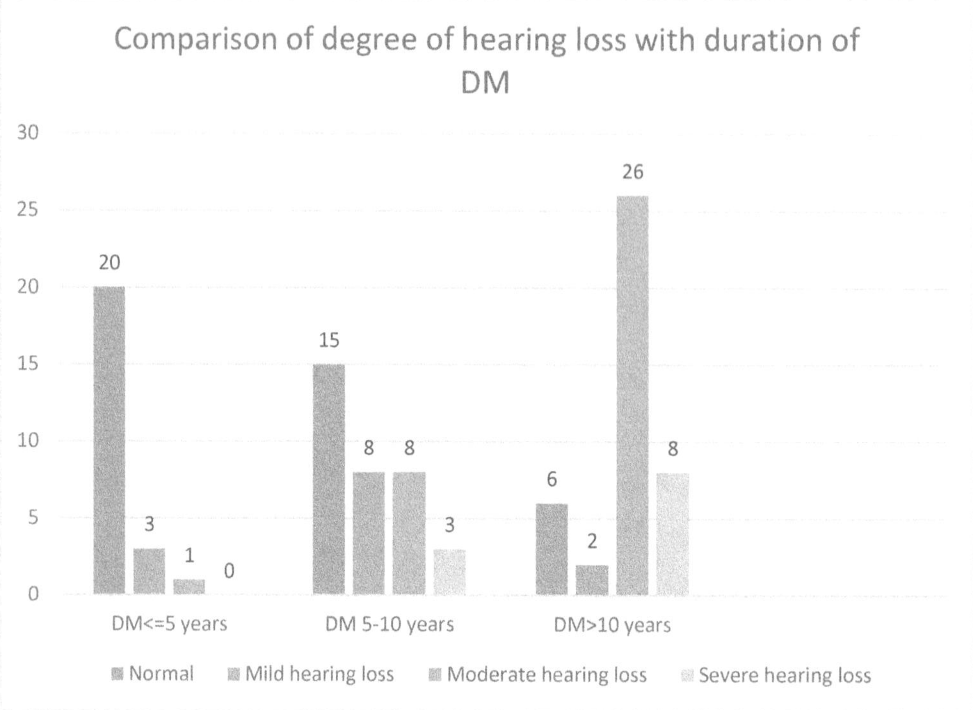

DISTRIBUTION BASED ON HbA1C VALUES:

Among the study group, 9 patients had their HbA1C value<7, 51 patients had their HbA1C value between 7 to 8 and 40 patients had their HbA1C value>8.

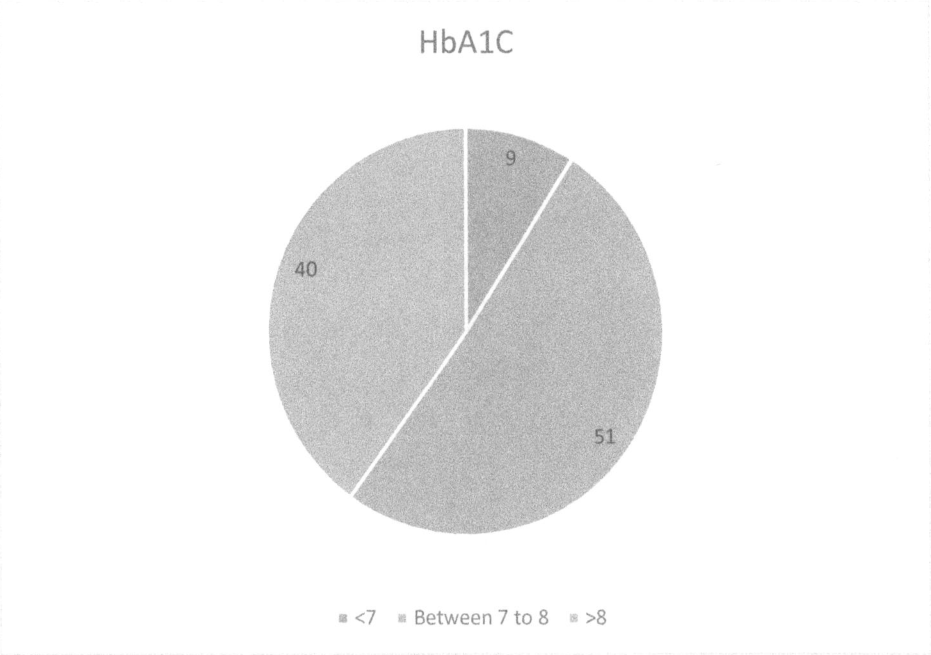

DISTRIBUTION OF HEARING LOSS BASED ON HbA1C VALUES:

Among the 9 patients whose HbA1C values are less than 7, 1 patients (11%) had hearing loss and 8 patients (89%) had no hearing loss. Among the 51 patients with HbA1C values between 7 and 8, 21 patients (41%) had hearing loss and 30 patients (59%) had no hearing loss. Among the 40 patients whose HbA1C values are more than 8, 37 patients (92.5%) had hearing loss and 3 patients (7.5%) had no hearing loss. The 'p' value is <0.05, making it statistically significant.

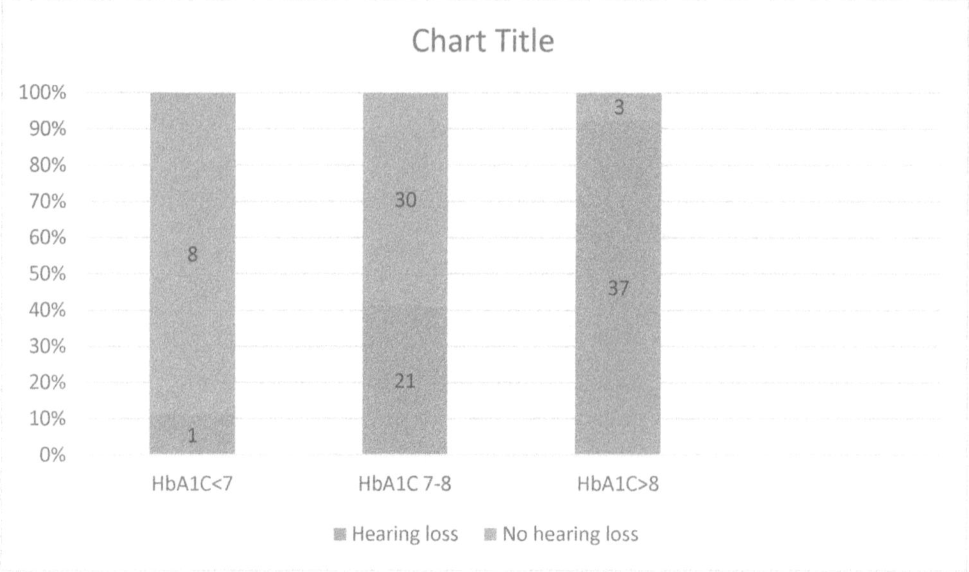

COMPARISON OF TYPES OF HEARING LOSS ACCORDING TO THEIR HbA1C VALUES:

Among the 9 patients whose HbA1C value is <7, 8 patients (89%) had no hearing loss and only 1 patient (11%) had sensorineural hearing loss. Among the 51 patients whose HbA1C is between 7 and 8, 4 patients (8%) had conductive hearing loss, 4 patients (8%) had mixed hearing loss, 13 patients (25%) had sensorineural hearing loss and 30 patients (59%) had hearing within normal limits. Among the 40 patients whose HbA1C is more than 8, 1 patient (2.5%) had conductive hearing loss, 3 patients (7.5%) had mixed hearing loss, 33 patients (82.5%) had sensorineural hearing loss and 3 patients (7.5%) had no hearing loss.

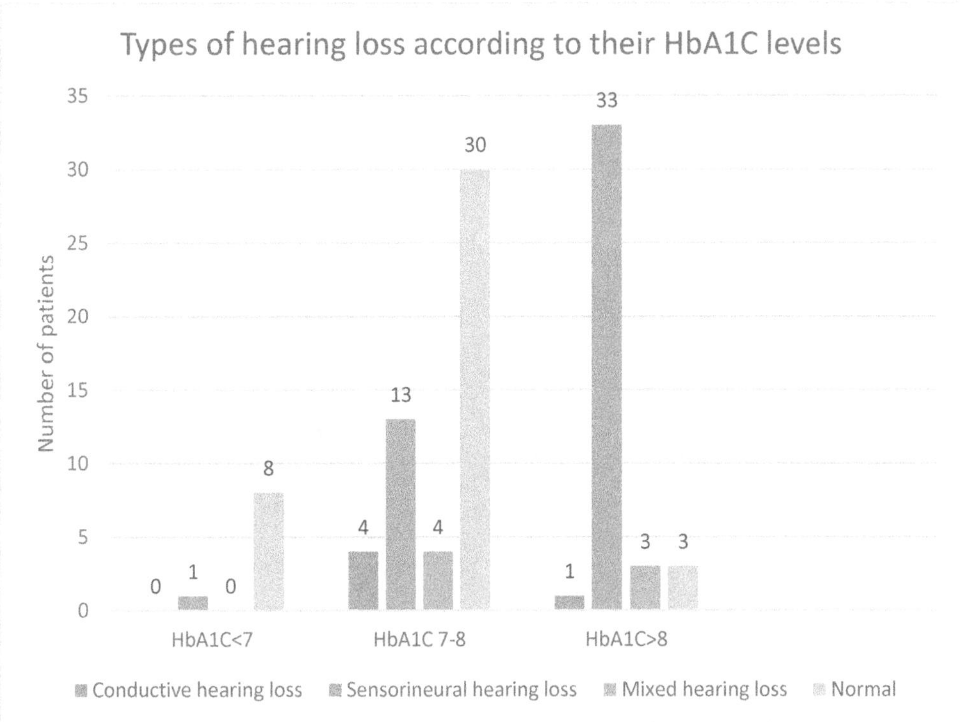

COMPARISON OF DEGREE OF HEARING LOSS AND HbA1C LEVELS:

Among the 9 patients whose HbA1C is <7, only 1 patient had mild hearing loss(11%) .Among the 51 patients whose HbA1C levels are between 7-8, 7 patients(14%) had mild hearing loss,14 patients(28%) had moderate hearing loss and 30 patients(58%) had hearing within normal limits .Among the 40 patients whose HbA1C levels are >8 ,5 patients(12.5%) had mild hearing loss,21 patients(52.5%) had moderate hearing loss,11 patients(27.5%) had severe hearing loss and 3 patients(7.5%) had hearing within normal limit. The 'p' value is <0.05, making it statistically significant.

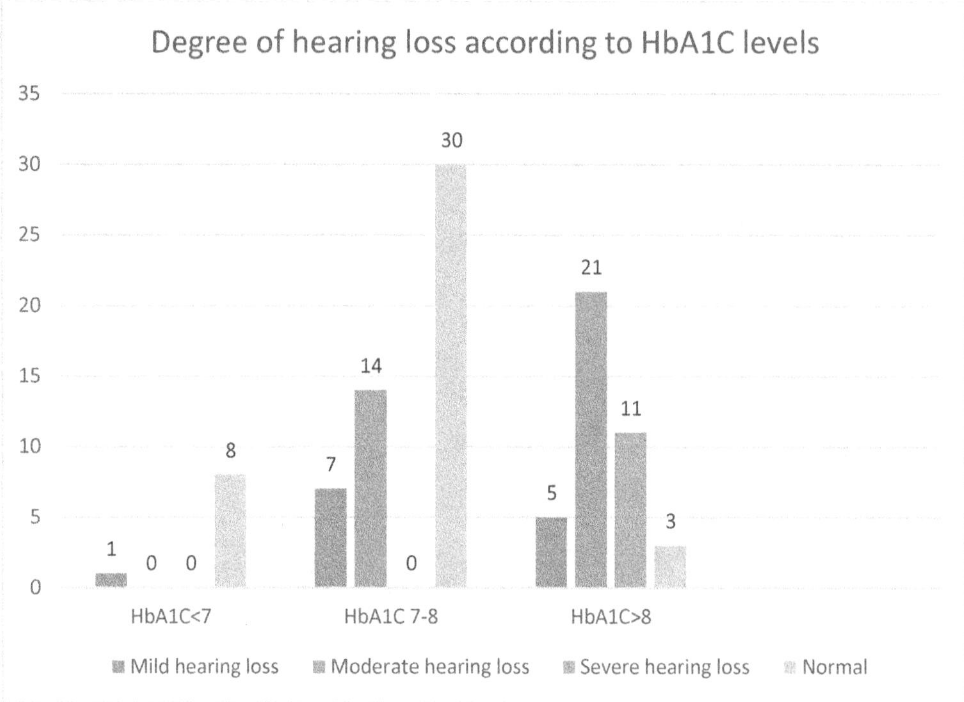
Degree of hearing loss according to HbA1C levels
35
30
25
20
15
10
5
0
1
0
0
8
7
14
0
30
5
21
11
3
HbA1C<7
HbA1C 7-8
HbA1C>8
Mild hearing loss
Moderate hearing loss
Severe hearing loss
Normal

DISCUSSION

This study shows the effect of diabetes mellitus on hearing loss with relation to age, gender, duration of disease, the type of hearing loss, degree of hearing loss and the correlation between HbA1C levels with the degree of hearing loss.

100 patients were included in the study and 51 among them were males and 49 were females. Majority of the study population (57%) were in the age group of 41-50 years.

Clinical examination that aided to maintain inclusion criteria were local examination of ear, nose, throat and systemic examination for other co-morbidities and the tuning fork tests.

After assessing and confirming the type and degree of hearing loss by PTA, confirmation and classification of degree of hearing loss was done with the help of WHO grades of hearing impairment [51].

AGE:

This study included patients in the age group of 31-55 years. The mean age group being 45.96 years and the maximum number of patients being in the age group of 46-50 years.

A similar study conducted by Rajendran S , Anandhalakshmi et al[52] also showed maximum incidence between 40- 50 years of age. Acquired sensorineural hearing loss can occur early in diabetic patients.

Usually age related sensorineural deafness is more common in people more than 60 years of age. But their study showed that diabetic patients are at higher risk of developing sensorineural hearing loss at a much younger age groups.

Another similar study conducted by Prathula Sivakumar et al showed maximum incidence of hearing loss in diabetic patients in the age group of 50-53 years.

GENDER:

100 patients were included in this study. 51 were males and 49 were females. From this study, we conclude that sex of an individual is not statistically significant ($p>0.05$) parameter in considering the susceptibility to develop hearing loss.

But according to Cullen and Cinnamond, diabetic male patients had hearing loss worse than diabetic female patients.

PREVALENCE OF HEARING LOSS:

This study showed 59% of the diabetic patients had hearing loss. Similarly, Friedman et al observed an incidence of 55% hearing loss in patients with diabetes. Weng et al(8) observed 67 patients with diabetes and found 44.8% among them had hearing loss which is lesser comparing to this study.

PREVALENCE OF HEARING LOSS WITH DURATION OF DIABETES MELLITUS:

24% of the patients in this study had diabetes mellitus for a period of <=5 years out of which 17% of them had hearing loss. 34% of the patients had diabetes mellitus between 5 -10 years out of which 56% of them had hearing loss. 42% of the patients had diabetes mellitus for >10 years out of which 86% of them had hearing loss.

In this study, it is seen that duration of diabetes has a positive correlation on the prevalence of hearing loss.

A study by Celik et al(53) noted that as diabetes duration raised to more than 15 years, the hearing loss incidence also increased. The influence on hearing loss was not significant after 15 years. This change was not observed in this study as cases in the study were all mostly found to be less than 15 years.

A similar study conducted by Nagaraj et al[54] noted that out of the 102 diabetic patients 12 patients(12%) were newly diagnosed diabetics, 43 patients(42%) had diabetes for less than 5 years, 38 patients(37%) had diabetes between 5-10 years and 9 patients(9%) had diabetes for a duration of >=10 years. The patients who had diabetes for more than 10 years are very less in this study when compared with our study. The same study showed that in the newly diagnosed diabetic patients,50% of the patients had normal hearing whereas 50% of the patients had hearing loss. In patients who had diabetes mellitus for more than 5 years,94% had hearing loss and 6% of them had normal hearing.

COMPARISON OF SEVERITY OF HEARING LOSS WITH THE DURATION OF DIABETES MELLITUS:

Among the 24 patients who had DM<=5 years ,20 patients (83%) had hearing within normal limits ,3 patients (13%) had mild hearing loss ,1 patient (4%) had moderate hearing loss. Among the 34 patients who had DM for 5-10 years ,8 patients (24%) had mild hearing loss ,8 patients (24%) had moderate hearing loss and 3 patients (8%) had severe hearing loss and 15 patients (44%) had no hearing loss. Among the 42 patients who had DM>10 years ,2 patients(5%) had mild hearing loss ,26 patients(62%) had moderate hearing loss ,8 patients(19%) had severe hearing loss and 6 patients(14%) had no hearing loss.

In a study conducted by Dayanand et al[55] ,in patients who had diabetes mellitus for 3-5 years , 24% of them had mild hearing loss ,4% of them had moderate hearing loss ,2% of them had moderately severe hearing loss and the remaining 70% people had normal hearing. Whereas in patients who had diabetes>5 years only 17% of them had hearing loss which is less compared to our study.

In a study conducted by Tiwari and Mudhol[56] ,in patients who had diabetes mellitus for 3-5 years ,60% of them had mild hearing loss ,27% of them had moderate hearing loss and 13% of them had moderately severe hearing loss. In patients who had diabetes for 6-9 years ,33% of them had mild hearing loss ,39% of them had moderate hearing loss ,18% of them had moderately severe and 10% of them had severe hearing loss. In patients who had diabetes for >=10 years ,40% of them had mild hearing loss ,33% of them had moderate hearing loss ,13% of them had moderately severe hearing loss and 13% of them had severe hearing loss.

COMPARISON OF TYPES OF HEARING LOSS WITH DURATION OF DIABETES MELLITUS:

Among the 24 patients whose duration of diabetes mellitus is <=5 years, 1 patient (4%) had conductive hearing loss, 2 patients (8%) had sensorineural hearing loss, 1 patient (4%) had mixed hearing loss and 20 patients (83%) had no hearing loss. Among the 34 patients who had diabetes mellitus for 5-10 years, 2 patients (6%)

had conductive hearing loss ,14 patients (41%) had sensorineural hearing loss and 3 patients (9%) had mixed hearing loss and 15 patients (44%) had no hearing loss. Among the 42 patients who had diabetes mellitus >10 years ,2 patients (5%) had conductive hearing loss ,31 patients (74%) had sensorineural hearing loss ,3 patients (7%) had mixed hearing loss and 6 patients (14%) had no hearing loss.

A study conducted by Nagaraj et al showed in patients who are newly diagnosed diabetics, 50% had normal hearing,42% of them had sensorineural hearing loss and 8% of them had mixed hearing loss. In patients who had diabetes mellitus for more than 5 years,79% of them had sensorineural hearing loss,16% of them had mixed hearing loss and the remaining people had hearing within normal limits which is similar to this study.

In a study conducted by Tiwari and Mudhol, in patients who had diabetes mellitus for 3-5 years ,68% of them had sensorineural hearing loss and 32% of them had normal hearing. In patients who had diabetes for 6-9 years ,89% of them had sensorineural hearing loss and 11% of them had normal hearing .In patients who had diabetes for >=10 years ,84% of them had sensorineural hearing loss and 16% of them had normal hearing.

DISTRIBUTION BASED ON HbA1C VALUES:

In this study, 9 patients had their HbA1C value<7, 51 patients had their HbA1C value between 7 to 8 and 40 patients had their HbA1C value>8.

In a similar study conducted by Nagaraj et al ,18 patients had HbA1C values between 6.5-7.5 ,22 patients had their HbA1C values between 7.5-8.5 ,8 patients had their HbA1C values between 8.5-10 and 14 patients had their HbA1C values >10.

In another study conducted by Tiwari and Mudhol ,36% of the patients had their HbA1C values below 7 ,46% of the patients had their HbA1C values between 7-9 and 18% of them had their HbA1C values >9.

HEARING LOSS AND HbA1C-CORRELATION:

In this study ,11% of the patients whose HbA1C levels are less than 7 had hearing loss. 41% of the patients had hearing loss whose HbA1C is between 7-8 .92.5% of the patients had hearing loss whose HbA1C is more than 8.

In a study conducted by Ooley et al. in 2017, it was revealed that in 175 patients with diabetic retinopathy and whose HbA1c value was known, there was a relationship between the hearing loss and the HbA1c value, as well as the degree of diabetic retinopathy.

Similarly, in another study conducted by Nagahama et al. in 2018, the hearing test results of the patient group with an HbA1c value of 7.3 and above revealed significantly higher losses at higher frequencies than other groups.

COMPARISON OF TYPES OF HEARING LOSS ACCORDING TO THEIR HbA1C VALUES:

Among the 9 patients whose HbA1C value is <7, 8 patients (89%) had no hearing loss and only 1 patient (11%) had sensorineural hearing loss. Among the 51 patients whose HbA1C is between 7 and 8, 4 patients (8%) had conductive hearing loss, 4 patients (8%) had mixed hearing loss, 13 patients (25%) had sensorineural hearing loss and 30 patients (59%) had hearing within normal limits. Among the 40 patients whose HbA1C is more than 8, 1 patient (2.5%) had conductive hearing loss, 3 patients (7.5%) had mixed hearing loss, 33 patients (82.5%) had sensorineural hearing loss and 3 patients (7.5%) had no hearing loss.

In a similar study conducted by Nagaraj et al, in patients who had their HbA1C values between 6.5-7.5, 67% of them had sensorineural hearing loss ,22% of them had mixed hearing loss and 11% of them had normal hearing. In patients had their HbA1C values between 7.5-8.5 ,88% of them had sensorineural hearing loss,9% had mixed hearing loss and 3% of them had hearing within normal limits. In patients had their HbA1C values>8.5 ,92% had sensorineural hearing loss and 8% had normal hearing which is almost similar to our study results.

In a similar study conducted by Tiwari and Mudhol, in patients who had their HbA1C values below 7 ,69% of them had sensorineural hearing loss, in patients who had their HbA1C values between 7-9 ,79% of them had sensorineural hearing loss and in patients who had their HbA1C values>9 ,87% of them had sensorineural hearing loss.

COMPARISON OF DEGREE OF HEARING LOSS AND HbA1C LEVELS:

Among the 9 patients whose HbA1C is <7, only 1 patient had mild hearing loss(11%) .Among the 51 patients whose HbA1C levels are between 7-8, 7 patients(14%) had mild hearing loss,14 patients(28%) had moderate hearing loss and 30 patients(58%) had hearing within normal limits .Among the 40 patients whose HbA1C levels are >8 ,5 patients(12.5%) had mild hearing loss,21 patients(52.5%) had moderate hearing loss,11 patients(27.5%) had severe hearing loss and 3 patients(7.5%) had hearing within normal limit.

In a similar study conducted by Nagaraj et al, in patients whose HbA1C levels are between 6.5-7.5 ,39% of them had mild hearing loss ,43% of them had moderate hearing loss ,7% of them had severe hearing loss and 11% of them had normal hearing. In patients whose HbA1C levels are between 7.5-8.5 ,25% of them had mild hearing loss ,66% of them had moderate hearing loss ,6% of them had severe hearing loss and 3% of them had normal hearing. In patients whose HbA1C levels are >8.5 ,8% of them had mild hearing loss ,42% of them had moderate hearing loss ,42% of them had severe hearing loss and 8% of them had normal hearing.

In a similar study conducted by Tiwari and Mudhol, in patients who had their HbA1C values below 7 ,84% of them had mild hearing loss and 16% of them had moderate hearing loss.In patients who had their HbA1C values between 7-9 ,31% had mild hearing loss ,51% of them had moderate hearing loss ,16% of them had moderately severe hearing loss and 2% of them had severe hearing loss. In patients who had their HbA1C values>9 ,30% of them had mild hearing loss ,15% of them moderate hearing loss ,35% of them had moderately severe hearing loss and 20% of them had severe hearing loss.

CONCLUSION

In summary, Hearing loss in type II diabetics showed no significant relation to gender. But early diabetics were prevalent as age groups were taken with consideration to their Type 2 DM duration.

Lastly, all associations lead to confirming significant traits of SNHL even though normal and mixed hearing loss components are insignificantly present among confirmed diabetic subjects.

This study gives significant insight into clinical outcome of diabetes on SNHL and the correlation between HbA1C levels and the severity of hearing loss.

Hearing loss observed in diabetics is mostly bilaterally symmetrical, progressive, ranging from mild to severe. Factors such as gender, family history, weight, occupation had no role in etiology of hearing loss. In addition to above factors, non communicable conditions or diseases, such as smoking, drinking, hypertension and thyroid disorders which may affect hearing loss in different manners were considered and excluded, so as to show direct effect of diabetes on hearing loss.

Even though clinical tests may help finding and proving hearing loss type and to a certain extent the degree as well, health care providers must take into account, the condition progression and refer to audiology for proper screening and evaluation for all diabetes mellitus cases, to have a better insight as well as preparedness for the outcome.

In future, there is a lot of scope to understand other aspects of SNHL with timely assessment of hearing threshold and diabetic management to improve auditory status.

REFERENCES

1) Bhaskar KN, Chalihadan S, Vaswani R, Rehaman CP. Clinical and Audiometric Assessment of Hearing Loss in Diabetes Mellitus. Int J Sci Stud 2014; 2(4):1- 16. https://www.ijss-sn.com/uploads/2/0/1/5/20153321/ijss_july-01.pdf

2) B.Viswanatha, V. Priyadarshini, M.S. Vijayashree, R. Rasika and Nisha Krishna; Effect of Type II Diabetes Mellitus on the 8th Cranial Nerve: A Prospective Study. Journal of Neurology and Epidemiology, 2014, 2, 25-29.

3) Gareth Williams, John C ,Pick up ,Textbook of diabetes mellitus ,Blackwell publishing Ltd , Oxford,Third edition,2003.

4) Cullen R ,Cinnamond N.J. "Hearing in diabetes" ,Journal of laryngology and otology,1993;107:179-182.

5) Virteniemi J et al , "Hearing thresholds in insulin dependent diabetes mellitus", Journal of laryngology and otology,1994;108:837-841.

6) Mohan V, Sandeep S, Deepa R, Shah B, Varghese C. Epidemiology of type 2 diabetes: Indian scenario. Indian JMed Res 2007: 217-30. https://www.ncbi.nlm.nih.gov/pubmed/17496352

7) Elzouki, Abdelaziz Y (2012). Textbook of clinical pediatrics (2 ed.). Berlin: Springer. p. 602. ISBN 9783642022012. https://en.wikipedia.org/wiki/ Hearing_loss#Classification

8) Weng SF, Chen YS, Hsu CJ, Tseng FY. Clinical features of sudden sensorineural hearing loss in diabetic patients. The Laryngoscope. 2005 Sep; 115(9):1676-80. DOI:10.1097/01.mlg.0000184790.91675.e3

9) http://www.who.int/en/news-room/fact-sheets/detail/deafness-and-hearing-loss.

10) Schnell O, Crocker JB, Weng J: Impact of HbA1c testing at point of care on diabetes management. J Diabetes Sci Technol. 2017, 11:611-617. 10.1177/1932296816678263.

11) Friedman SA, Schulman RH, Weiss S. Hearing and diabetic neuropathy. Arch Intern Med. 1975; 135: 573–6.

12) Lasisi OA, Nwaorgu OG, Bella AF. Cochleovestibular complications of diabetes mellitus in Ibedan, Nigeria. Int Congr Series. 2003; 1240: 1325–8.

13) Taylor IG, Irwin J. Some audiological aspects of diabetes mellitus. J Laryngol Otol. 1978; 92: 99–113.

14) Kakarlapudi V, Sawyer R, Staecker H. The effect of diabetes on sensorineural hearing loss. Otol Neurotol. 2003; 24: 382–6.

15) Axelson A, Fagerberg SE. Auditory functions in diabetics. Acta Otolarngol. 1968; 66: 49–63.

16) Carmen RE, Svihovec D, Gocka EF, Ermshar CB, Gay GC, Vanore JF, et al. Audiometric configuration as a reflection of diabetes. Am j Otol. 1988; 9: 327–33.

17) Kurien M, Thomas K, Bhanu TS. Hearing thresholds in patients with diabetes mellitus. J Laryngol Otol. 1989; 103: 164–8.

18) Dalton SD, Cruickshanks KJ, Klein R, Klein BE, Wiley TL. Association of NIDDM and hearing loss. Diabetes Care. 1998; 21: 1540–4.

19) Wilson WR, Laird N, Moo-Young G, Soeldner JS, Kavesh DA, MacMeel JW. The relationship of idiopathic sudden hearing loss to diabetes mellitus. Laryngoscope. 1982; 92: 155–60.

20) Zelenka J, Kozak P. Disorders in blood supply of the inner ear as early symptom of diabetic angiopathy. J Laryngol Otol. 1965; 79: 314–9.

21) Salvinelli F, Miele A, Casale M, Greco F, D'Ascanio L, Firrisi L, et al. Hearing thresholds in patients with diabetes. Int J Otorhinolaryngol 2004 ;3(1). DOI: 10.5580/252e.

22) Gupta S, Eavey RD, Wang M, Curhan SG, Curhan GC: Type 2 diabetes and the risk of incident hearing loss. Diabetologia. 2019, 62:281-285. 10.1007/s00125-018-4766-0 15.

23) Ren H, Wang Z, Mao Z, Zhang P, Wang C, Liu A, Yuan G: Hearing loss in type 2 diabetes in association with diabetic neuropathy. Arch Med Res. 2017, 48:631-637.

24) Cohen Atsmoni S, Brener A, Roth Y: Diabetes in the practice of otolaryngology . Diabetes Metab Syndr. 2019, 13:1141-1150

25) 25.Karabulut H., Karabulut I., Dağli M., Bayazit Y.A., Bilen S., Azdin Y., et. al.: Evaluation of outer hair cell function and medial olivocochlear efferent system in patients with type II diabetes mellitus. Turk J Med Sci 2014; 44: pp. 150-156.

26) Sunkum A.J., Pingile S.: A clinical study of audiological profile in diabetes mellitus patients. Eur Arch Oto-Rhino-Laryngol 2013; 270: pp. 875-879.

27) Misra V., Agarwal C.G., Bhatia N., Shukla G.K.: Sensorineural deafness in patients of type 2 diabetes mellitus in Uttar Pradesh: a pilot study. Indian J Otolaryngol Head Neck Surg 2013; 65: pp. 532-536.

28) Kakarlapudi V., Sawyer R., Staecker H.: The effect of diabetes on sensorineural hearing loss. Otol Neurotol 2003; 24: pp. 382-386.

29) Akinpelu O.V., Mujica-Mota M., Daniel S.J.: Is type 2 diabetes mellitus associated with alterations in hearing? A systematic review and meta-analysis. Laryngoscope 2014; 124: pp. 767-776.

30) Bainbridge K.E., Hoffman H.J., Cowie C.C.: Risk factors for hearing impairment among U.S. adults with diabetes:National Health and Nutrition Examination Survey 1999-2004. Diabetes Care 2011; 34: pp. 1540-1545.

31) Kidd Iii AR, Bao J: Recent advances in the study of age-related hearing loss: a mini-review . Gerontology. 2012, 58:490-496. 10.1159/000338588

32) Palomar García V, Abdulghani Martínez F, Bodet Agustí E, Andreu Mencía L, Palomar Asenjo V: Drug-induced otoxicity: current status . Acta Otolaryngol. 2001, 121:569-572.

33) Sliwinska-Kowalska M, Davis A: Noise-induced hearing loss. Noise Health. 2012, 14:274-80.

34) Bainbridge KE, Hoffman HJ, Cowie CC. Diabetes and hearing impairment in the United States: audiometric evidence from the National Health and Nutrition Examination Survey, 1999 to 2004. Annals of internal medicine. 2008 Jul 1; 149(1):1-0.

35) Bener A, Salahaldin AH, Darwish SM, Al-Hamaq AO, Gansan L. Association between hearing loss and type 2 diabetes mellitus in elderly people in a newly developed society. Biomedical research. 2008; 19(3). https://www.researchgate.net/publication/228742760

36) Ren J, Zhao P, Chen L, Xu A, Brown SN, Xiao X. Hearing loss in middle-aged subjects with type 2 diabetes mellitus. Archives of medical research. 2009 Jan 1;40(1):18-23. https://doi.org/10.1016/j.arcmed.2008.10.003

37) Bamanie AH, Al-Noury KI. Prevalence of hearing loss among Saudi type 2 diabetic patients. Saudi medical journal. 2011; 32(3):271-4. https://www.ncbi.nlm.nih.gov/pubmed/21384063.

38) B.Viswanatha, V. Priyadarshini, M.S. Vijayashree, R. Rasika and Nisha Krishna; Effect of Type II Diabetes Mellitus on the 8th Cranial Nerve: A Prospective Study. Journal of Neurology and Epidemiology, 2014, 2, 25-29.

39) Pandey D, Pandit A, Pandey AK. Study of audio vestibular dysfunction in type2 diabetes mellitus. IAIM, 2016; 3(6): 23-26.

40) Louise Ziani de David , Marcele Machado Finamor, Ceres Buss did a study in 2012 on Possible hearing implications of diabetes mellitus: a literature review. Rev. CEFAC. 2015 NovDez; 17(6):20182023.

41) Bhaskar KN, Chalihadan S, Vaswani R, Rehaman CP. Clinical and Audiometric Assessment of Hearing Loss in Diabetes Mellitus. Int J Sci Stud 2014; 2(4):1- 16.

42) Meena R, Sonkhya D, Sonkhya N. Evaluation of hearing loss in patients with type 2 diabetes mellitus. International Journal of Research in Medical Sciences. 2017 Jan 2; 4(6):2281-7. http://dx.doi.org/10.18203/2320-6012.ijrms20161800.

43) Ooley C, Jun W, Le K, et al.: Correlational study of diabetic retinopathy and hearing loss . Optom Vis Sci. 2017, 94:339-344

44) Nagahama S, Kashino I, Hu H, et al.: Haemoglobin A1c and hearing impairment: longitudinal analysis using a large occupational health check-up data of Japan. BMJ Open. 2018, 8:e023220.

45) Anthony Wright, "Anatomy and Ultrastructure of human ear", in Scott- Brown's Otolaryngology, 1997, 6th edition, Vol. 1, 1/1-49.

46) Cummings Otolaryngology Head & Neck Surgery, Volume – 3. Edition – 5, Mosby Elsiever, USA, 2010; Part 7 and 8, Section 1, 2, 5, 7.

47) Scott Brown's Otorhinolaryngology, Head and Neck Surgery, Volume – 3, Edition - 7, CRC PRESS, UK, 2008; Chapters 226

48) P L Dhingra, Shruti Dhingra, Diseases of Ear, Nose and Throat, 6th Edition, Elsiever, India, 2014; Section 1,Chapter 1,2.

49) Bredberg G: Cellular and nerve supply of the human organ of corti, Acta Otolaryngol Supl 236:1-135, 1968. Occupational Safety and Health Administration, Department of Labor: Occupational noise exposure: hearing conservation amendment. Fed Reg 1981; 46:4078-4179.

50) James O. Pickles, "Physiology of Hearing", in Scott-Brown's Otolaryngology, 1997, 6th edition, Vol.1, 2/1-31.

51) Donald Henderson, Bhoua Hu, Pattern and mechanism of noise induced cochlear pathology.

52) Rajendran S , Anandhalakshmi , Mythili B , Viswanatha Rao. Evaluation of the Incidence of Sensorineural hearing loss in Patients with Type 2 Diabetes Mellitus. Int J Biol Med Res. 2011; 2(4): 982 – 987.

53) Celik O, Yalçin S, Celebi H, Oztürk A. Hearing loss in insulin dependent diabetes mellitus. Auris Nasus Larynx.1996;23:127-32.

54) Dr Nagaraj Bangalore Thimmasettaiah, Dr Ravi Shankar, Dr Ravi G C, Dr somasundar Reddy. A one year prospective study of hearing loss in diabetes in general population. Translational biomedicine Vol. 3 No. 2:1 doi: 10.3823/433.

55) Dayanand, J. Dheebika, S. Prathula and S. Palaninathan. A Study on Prevalence of Hearing Loss as a Complication of Diabetes. DOI: http://dx.doi.org/10.24018/ejmed.2020.2.4.40.

56) Aniruddh Tiwari, Ramesh S. Mudhol. Prevalence of sensorineural hearing loss among type-II diabetes mellitus patients attending KLES Dr. Prabhakar Kore Hospital and MRC: A cross-sectional study. DOI: 10.4103/kleuhsj.kleuhsj_187_17.

www.ingramcontent.com/pod-product-compliance
Lightning Source LLC
LaVergne TN
LVHW082357200926
848481LV00012B/3593
9798231749706